VELOCIPEDOMANIA

VELOCIPEDOMANIA

A CULTURAL HISTORY OF THE VELOCIPEDE IN FRANCE

CURRY CROPPER

AND

SETH WHIDDEN

LEWISBURG, PENNSYLVANIA

Library of Congress Cataloging-in-Publication Data

Names: Cropper, Corry, author. | Whidden, Seth Adam, 1969- author.
Title: Velocipedomania : a cultural history of the velocipede in France / Corry Cropper and Seth Whidden.
Description: Lewisburg, Pennsylvania : Bucknell University Press, [2023] | Includes bibliographical references and index.
Identifiers: LCCN 2022009352 | ISBN 9781684484331 (paperback) | ISBN 9781684484348 (hardback) | ISBN 9781684484355 (epub) | ISBN 9781684484379 (pdf)
Subjects: LCSH: Velocipedes—France—History—19th century. | Velocipedes—History—Sources. | Popular culture—France—History—19th century. | Cycling—France—History—19th century.
Classification: LCC TL405 .C76 2023 | DDC 629.227/20944—dc23/eng/20220803
LC record available at https://lccn.loc.gov/2022009352

A British Cataloging-in-Publication record for this book is available from the British Library.

∞ The paper used in this publication meets the requirements of the American National Standard for Information Sciences—Permanence of Paper for Printed Library Materials, ANSI Z39.48-1992.

www.bucknelluniversitypress.org

Distributed worldwide by Rutgers University Press

Manufactured in the United States of America

CONTENTS

VELOCIPEDOMANIA

INTRODUCTION

VELOCIPEDOMANIA

> The time has come to speak of the velocipede. The velocipede is in our culture, it has entered our blood, it has become an institution. . . . The velocipede had to be born. It is now synonymous with our time and with our culture.
>
> —*Charles Yriarte, 1869*

This work is an homage to folly, a tribute to a period of excessive exuberance for a machine that inspired music, plays, and novels—or, more abstractly, new hymns, liturgies, and sacred texts. Almanacs were reoriented by the fervor that spread across France; disciples were trained in newly formed schools and by manuals that stipulated how to speak about, dress for, and use this new machine; and the national zeal gave rise to a set of values that promised new and unexpected relationships, freedom of movement, a way around social and economic barriers, and improved physical health. This book is about a craze that swept through France in the late 1860s, as recorded in training manuals, scripts, songs, and newspaper articles, that both reflected and contributed to reshaping French society: *velocipedomania*. At the nexus of the scientific philosophy of positivism and democratic ideals on which the Third Republic would be founded, the new machine offered an early glimpse of how its 1890s successor, the bicycle, would contribute to the continuing evolution of French society.

This volume examines a unique and narrow window of time (1868–69) in which the ardor for the velocipede caused a literal mania in France. Though the Franco-Prussian War and dark days of 1871 would put a sudden end to the enthusiasm for the velocipede, velocipedomania's cultural imprint would eventually give rise to France's great national symbol, the *vélo,* to the Tour de France, and to a form of worship that has spread well beyond France's borders and is now practiced the world over.[1] We have brought together translations and analyses of works that embody velocipedomania: the anonymous 1868 *Note sur le vélocipède à pédales et à frein de M. Michaux par un amateur* (hereafter *Note on Monsieur Michaux's Velocipede*); a short operetta staged that same year by Henri Blondeau titled *Dagobert et son vélocipède* (hereafter *Dagobert and His Velocipede*); the most substantial and influential work on the velocipede, the 1869 *Manuel du vélocipède* (hereafter *Manual of the Velocipede*), compiled by Le Grand Jacques (pseudonym of Richard Lesclide); and a short sampling of poetry about the velocipede.[2] Our aim is to show the social impact of the velocipede, to examine how it reflected the French social imaginary, and to explore how it became embedded in French culture.[3]

While focusing on the velocipede's cultural history, we do not mean to offer a broad history that situates it in the long evolution of two-wheelers. Readers seeking such historical context will find it in David Herlihy's book *Bicycle: The History* (2004), which has an excellent chapter on the emergence of the velocipede. Paul Smethurst's *The Bicycle: Towards a Global History* (2015) also examines the historical emergence of the velocipede in its early pages. Keizo Kobayashi's *Histoire du vélocipède de Drais à Michaux 1817–1870: Mythes et réalités* (1991) is the most thorough history of the invention, production, and evolution of the velocipede. And Tony Hadland and Hans-Erhard Lessing's *Bicycle Design: An Illustrated History* (2014) is the definitive resource on patents, nomenclature, and the development of everything from skates to sewing machines to velocipedes and bicycles of all sorts. Our task is not to compete with these historians' excellent work; rather, it is to build on it and to study the cultural manifestations and dissemination of the velocipede and the mania it inspired.[4]

THE VELOCIPEDE'S FIRST PEDAL STROKES

The velocipede (later called a "boneshaker" by English speakers) was an iron-framed, wooden-wheeled machine. Pierre Michaux, who began manufacturing velocipedes in the 1860s (the first ads appeared in 1867), engineered pedals directly to the front axle, added a brake for the rear wheel, and built a frame that provided some amount of shock absorption. Though this origin story would be contested by Pierre Lallement, René Olivier, and others, Michaux's company became the primary manufacturer and his name became synonymous with the new machine.[5] Most models also included what looked like a fender over the front wheel: a lantern could be affixed to this attachment (though it was more often affixed to the handlebars), and flanges on its side served as footrests while the velocipedist (or *vélocipédeur, véloceman*, or *écuyer* [horseman or jockey]—the term varied) coasted downhill and the pedals spun frenetically. Michaux's company also offered riding lessons from its storefront near the Champs-Elysées in Paris.

One of the earliest descriptions of the velocipede appeared in July 1867 in the biweekly newspaper *Le Sport*, a publication founded by—and frequently including articles from—prolific writer Eugène Chapus, who also published a number of manuals on etiquette, fashion, and the lifestyle of the upper class under his pseudonym, the Viscount de Marennes.[6] In an article titled "En véloce! En véloce!" the journalist G. d'O explains, "This is the call to assembly that several intrepid Parisians, fanatics of this new form of locomotion, have loudly repeated of late. This shout first rang out from the avenue Montaigne [the location of Michaux's workshop] then traveled to the Exposition universelle with the speed of a velocipede; and it will be from the Champ-de-Mars [the venue for many of the Exposition's exhibits] that this cry will spread like a flash, traveling with each exhibitor to the ends of the earth."[7] After describing the velocipede, the article turns its attention to people associated with it: Michaux is like a modern Prometheus—"The velocipede . . . is the work of Monsieur Michaux, a French mechanic who patented the attachment of pedals several years ago"[8]—and the people who have ordered and ridden one of his machines include a colonel, a doctor, a priest, and people from far-flung places like England and China.

> Clubs of *vélocipédeurs* are being organized and when we see approximately one thousand names of society's finest on the inventor's order list, we have no doubt these clubs will succeed. The *high-life* has taken this new exercise under its patronage, granting it a place in the world of sport alongside cricket, skating, and pigeon shooting. Wagers have already been placed on private races held in the Bois de Boulogne, and as soon as a club of *vélocipédeurs* is formally constituted with aristocratic members—like the shooting club and the ice-skating club—we will have brilliant races between gentlemen; we could already easily pick favorites.[9]

Given that the article ends with Michaux's address, the prices of his velocipedes, and information about free lessons, we suspect the article was paid for by Michaux or his promoters. Whatever the case, this article praising Michaux, the presence of Michaux velocipedes at the Exposition universelle, and Michaux's velocipede school near the Champs-Elysées cemented his place as the mythical creator of the new machine. Even as velocipedes were appearing in other countries, this article also situates France as the cultural home of the velocipede and hints that the French were the first to embrace, promote, and even gamble on velocipede races. Finally, Chapus's newspaper envisions the velocipede as a purveyor of French universalism: much like the Tour de France would later take civilization to rural France, the velocipede is destined to export French technology and culture "to the ends of the earth."[10]

PICKING UP SPEED

The velocipede proved so much more popular than its predecessors that within just a few short years, by 1870, there were more than forty velocipede clubs in France.[11] The velocipede succeeded beyond the *draisine*—a two-wheeled, wooden vehicle propelled by riders pushing the ground with their feet—for a number of reasons: first and foremost, the ability to use pedals meant that speed was no longer limited by the regular touching of foot to road surface, allowing riders to be more stable and to go farther and faster while more efficiently expending their energy.[12] Though it required defenders to make the case, as we will see in the *Note on Monsieur Michaux's Velocipede*, the new machine held the promise of practical uses for communication, travel, emergency services, and health. In addition, as argued in the article from *Le Sport* quoted earlier, riding the velocipede was a way to integrate the "high life" and to belong to a group of people who enjoyed progress, had the means to pay for a velocipede, and knew how to ride. Even then, the velocipede remained a crude instrument at best: only the strongest riders could go much over sixteen kilometers per hour, the suspension did little to dampen the shock of rough roads, the lack of gears often meant pushing the velocipede uphill, and, with inefficient brakes, going downhill could quickly turn to disaster. Most importantly for our study, though, the velocipede pedaled through the very heart of French culture; as journalist Charles Yriarte maintained in 1869, the velocipede "entered our blood. . . . It is now synonymous with our time and with our culture." In a relatively short time span it emerged as a ubiquitous marker of modernity, of freedom, and of Parisian, even French, identity.[13]

The popularity of the velocipede was attested in newspapers, treatises, and music. As early as February 1868, Léon Bienvenu (also known as Touchatout), writing in the weekly paper *L'Eclipse*, announced, "The velocipede is a true sign of the times. It has been all the rage this year; and I believe this enthusiasm is merely the prelude to an even greater era. . . . Once it is fully embedded in our culture, I believe the velocipede will prove incredibly useful. Each citizen will have one hanging in their entryway, and when they go out, they will take their velocipede along with their coat and hat."[14] In May of the same year, the *Industriel de Saint-Germain-en-Laye*, an industrial, agricultural, and administrative newspaper, noted the growing popularity of the velocipede: "Velocipedes continue to multiply in Paris. Soon everyone who can't afford two or four horses will have a velocipede. Some are already faster than horse-drawn carriages. Every day now, hundreds of these vehicles can be seen on the boulevards, the *quais*, the rue de Rivoli, the Champs-Elysées, and on the avenues of the Bois de Vincennes and the Bois de Boulogne."[15] The *Manual of the Velocipede*'s 1869 preface sums up the general feeling about the velocipede in these terms: "The popularity of the Velocipede is more than a trend or a sport; it is a fever."[16]

NAVIGATING OBSTACLES

Of course, the velocipede had its detractors, who, though they did not seem to have any particular political agenda, complained about it clogging streets and worried that it would dehumanize and emasculate Frenchmen or undermine French cultural superiority. An entire front page of *Le Petit Journal* (July 5, 1868) was dedicated to the velocipede. Written by Timothée Trimm, the article begins by acknowledging the commotion caused by the new machine: "It's the furor of the moment. It makes everyone stop on public roads. It's the preoccupation of coachmen and the worry of pedestrians. It can be found in the parks, on the boulevards, in the streets. We have to speak about it both as a sign of the times and as a manifestation of the tastes and the affections of the crowd. The velocipede is in fashion."[17] Trimm then describes the velocipede but maintains it has just one advantage over the horse: "People won't try to eat it."[18] Trimm continues:

> The velocipede has its critics. It is difficult to make it stop quickly when faced with an obstacle, to make it climb a hill or to descend. Some affirm that it is not wise to ride the streets of Paris on this new hobbyhorse. Others maintain that the velocipede only works well on smooth, gentle, paved, uniform roads. Some medical doctors maintain that the repetitive motion of the leg, constantly pushing on the wheels, is not healthy. When walking, the entire body is in

> movement. . . . The velocipede, as it exhausts the legs, leaves the rest of the body inert . . . thereby tiring part of the human machine while leaving the rest of the body in a state of dangerous immobility.[19]

But Trimm's greatest fear is that the velocipede will damage France's literary reputation. After mentioning velocipede races held at the imperial residence at Saint-Cloud at the end of May, he explains, "I don't want to be taken as an enemy of progress simply because I prefer beef to horse meat, and a four-wheeled vehicle hitched to a valiant and gentle steed to a well-built velocipede. But I just can't imagine Hugo, Lamartine, Augier, Autran, Viennet, Banville, Coppée, Daudet, the poets old and young, mounting a velocipede instead of the winged horse of Parnassus [Pegasus]. . . . [20] And I'll never get used to the idea of hearing Shakespeare's Richard III cry out in his bellicose zeal: 'A velocipede! My kingdom for a velocipede!'"[21] The tragedy of the velocipede in Trimm's eyes is that it could humiliate France's novelists and poets and deprive them of their glorious equestrian ride to the mountain of the gods! He imagines that this mechanical vehicle will replace noble chargers and make the great icons of French culture look ridiculous. Though it is seen as a negative here, as early as mid-1868 the velocipede is already viewed as synonymous with French culture.

In a similar vein, *Le Journal amusant* published a series of sketches by Alfred Grévin of a velocipedist in March 1869. The first is captioned "Au pas" (Walk), the second "Au trot" (Trot), and the third "Au Galop! Au Galop!! Au Galop!!!" (Gallop! Gallop!! Gallop!!!). "Au pas" depicts a man riding a three-wheeled velocipede through the Longchamps park with a woman and her dog perched behind. The woman is leisurely smoking a cigarette and the man is comfortably pedaling. In "Au trot," the cigarette is gone and the woman is raising a whip while tightening a leash around the rider's neck (see cover illustration). In the last illustration of the series (figure I.1), it is clear that the man has replaced the horse: he is hunched over the handlebars, his mouth open, eyes closed, giving maximum physical effort; his hat has blown off and trails behind. The female passenger now stands as if in a carriage, imperiously

A LONGCHAMPS, — par A. GRÉVIN (suite).
LES NOUVEAUX VÉLOCIPÈDES.
AU GALOP! AU GALOP!! AU GALOP!!!

FIGURE I.1. Alfred Grévin, "At Longchamps: The New Velocipedes," *Le Journal amusant*, March 27, 1869.

whipping the velocipedist, whom she controls via the reins she holds in her hand. Adding insult to injury, the small dog bites or sniffs the man from behind; the animal has a comfortable ride while the dehumanized velocipedist strains forward like a beast of burden. If the velocipede has replaced the horse, the rider has assumed the horse's work and finds himself humiliated and debased by the new machine.

In his aforementioned article in *L'Eclipse*, after noting the ubiquity of the velocipede, Léon Bienvenu cannot help but take a sarcastic swipe at its fans:

> It is evident that man now finds natural means of transportation to be out of fashion. And the dispiritingly monotonous way of getting from point A to point B up until now, putting the left leg in front of the right leg and then the right in front of the left, seems beneath him.
>
> He is looking for something else. . . .
>
> The velocipede presents tremendous advantages. First, it enables people to travel great distances without getting their feet in muck; only the wheels touch the mud and, thanks to their rotational movement, they cover the velocipedist from the bottom of his jacket to the top of his head; but never any higher.
>
> What's more, the velocipede's maintenance costs are very low in hilly regions, since riders must carry it on their back when going uphill, meaning there is almost no wear and tear.
>
> And finally, typical carriages are susceptible to wheel damage since the wheels are exposed. . . . Reparations can be exceedingly onerous. But with velocipedes, there is nothing to fear since your wheels are protected by your legs on both sides.[22]

To counter its detractors, the velocipede's early promoters sought to establish its utilitarian applications. Alexis-Georges Favre's 1868 brochure, with its descriptive title *Le Vélocipède, sa structure, ses accessoires indispensables, le moyen d'apprendre à s'en servir en une heure* (The Velocipede, Its Structure, Its Requisite Accessories, and How to Learn to Use It in One Hour), outlines the velocipede's utility, lists its accessories (reflectors, seats, shin-wraps, oil, etc.), and indicates prices.[23] It emphasizes the savings to be accrued by riding a velocipede rather than paying coach fare or buying feed for a horse, and it describes how velocipedes can save time and lead to more profit for businesses.[24] Favre, who took mail orders for velocipedes and accessories at his shop south of Lyon, emphasized the utility of the velocipede primarily in order to generate more business. In a short essay addressed to the Rouen velocipede club in 1869, doctor Élie Bellencontre argues in favor of the physical and emotional benefits of riding a velocipede. "Today," writes Dr. Bellencontre, "the velocipede can no longer be considered a simple toy. It is a utilitarian object, a means of locomotion, and I intend to demonstrate how, when applied to hygiene and exercise, it can support good health, provide pleasure, and how it leads to improving the good morality of the masses, an objective to which our intellect is always drawn."[25] *Le Manuel du véloceman* (The Manual of the Véloceman), penned by well-known architect Alfred Berruyer in Grenoble in 1869, begins with an optimistic declaration that draws on the lexical field shared by the equestrian and the velocipedist: "The *véloce* is now counted among the useful mounts of civilized peoples. Once its path is well traced on our roads, it will conquer the entire world."[26]

The most prominent argument in favor of the utility of the velocipede, a treatise that predicts many practical uses of the bicycle today—from delivery services to commuting—is the 1868 *Note on Monsieur Michaux's Velocipede* included in this volume. The anonymous author, a self-proclaimed "enthusiast" and an employee at the Ministry of the Navy, systematically outlines a variety of applications: promoting hygiene, improving communication, speeding up emergency response times, and patrolling vast areas more efficiently. Given the text's somewhat obsequious tone toward the state and its "auguste" sovereign—Napoleon III—it appears to have been written, at least in part, to convince various administrators to improve roads and to promote the use of velocipedes by government employees. It is the first example of public advocacy on behalf of two-wheeled human-powered vehicles.

ROLLING THROUGH FRENCH CULTURE

In addition to these practical treatises, the velocipede was also well represented in illustrations, songs, newspapers, and the theater. France's best-known nineteenth-century illustrator, Honoré Daumier, viewed the velocipede as the embodiment of 1868 France. Against the backdrop of rumors of war with Prussia, broken treaties, and a Franco-Prussian arms race, Daumier published "My Velocipede" in the infamous daily newspaper *Le Charivari* in September 1868 (figure I.2).

FIGURE I.2. Honoré Daumier, "My Velocipede," *Le Charivari*, September 17, 1868.

Published just two days before the Glorious Revolution in Spain (which Otto von Bismarck would turn to Prussia's advantage), Daumier's illustration depicts a feminine "Peace" wearing a loose-fitting dress floating in the breeze behind her as she speeds forward on her velocipede. However, the top of the velocipede has been replaced by a canon, suggesting that peace is tenuous and that France is heading ineluctably toward war. That the Marianne-like figure is riding from

FIGURE I.3. Honoré Daumier, "December 31: Table of Contents," *Le Charivari*, December 31, 1868.

left to right—from west to east—further implies that the danger is France's eastern neighbor, Prussia.[27]

Later that same year, Daumier turned again to the velocipede to grimly represent the year 1868 (figure I.3). In "December 31: Table of Contents," Daumier's female velocipedist, now a specter whose dress is in tatters, pedals past the headstones of those who have recently died: the composer Gioachino Rossini, politicians Pierre-Antoine Berryer and Léonor-Joseph Havin, and the novelist Félicien Mallefille, among

FIGURE I.4. Jules-Barthélemy Péaron, *Quadrille of the Velocipedes*, 1869.

others. Here, the velocipede stands as a dark symbol of contemporary France; but as in Daumier's "My Velocipede," the new machine is something of a double-edged sword: it portends freedom, peace, and mobility but coexists with war and death.

In a lighter vein, the cover page to an 1869 composition for piano, by the illustrator Jules-Barthélemy Péaron, features men and women in the appropriate pastoral setting and in the proper positions for the quadrille (a type of popular dance), but they are on velocipedes (figure I.4). The dancers-on-wheels seem to defy gravity, nonchalantly holding their poses without putting a foot on the ground, while musicians play under a tree in the background. It seems the composer, Eugène Baron, hoped to revitalize a traditional form of pastoral music by associating it with the velocipede, which the public would have quickly recognized as a symbol of modernity.[28] The velocipede was so popular that it gave rise to a musical parody written "in the patois of Lille" (a city in northern France) and sung to a tune from

"Le Postillon de Lonjumeau," an 1836 opéra comique by Adolphe Adam. The 1869 parody is about a velocipedist named Gaspard who, "like Don Quixote, is followed around by prostitutes when he rides."[29]

The publication *La Chanson illustrée* picked up on this theme of eroticism in its issue of May 2, 1869, which featured a group of velocipedists on its cover, led by a brilliantly dressed woman in tight-fitting trousers (figure I.5). The text on the bottom right of the cover points readers to the next page for the lyrics to a song by Alexandre Flan. It is titled "The Velocipedes" and includes instructions that it be sung to a tune composed by Paul Blaquière in 1865, "La Vénus aux carottes."[30] The velo-song's eight stanzas speak to the mania the velocipede inspired:

> A vehicle that is fashionable everywhere, . . .
> [the velocipede] is a fever without a cure. . . .
> Today its reign has come!
> In the Luxembourg Gardens or the Bois de Boulogne,
> Students, artisans, and willowy youth
> Ride their bicycles without shame,
> Quick as a flash, devouring the road.
> Even prostitutes give in to this fad.
> Look, for example, at the one Hadol [the illustrator] put on the cover
> On her velo, her velo, her velocipede. . . .
> How do we classify this new animal?
> Is it a beast or a machine?
> It is related to the sphynx as much as to the horse. . . .
> Postal workers want them! And have you heard
> That people are speaking of having the cavalry ride them?
> They also say, and this is news,
> That Henri IV is crazy about them, and he is asking
> That his horse on the Pont Neuf be replaced
> By a velo, a velo, a velocipede![31]

These lyrics emphasize recurring themes associated with the velocipede. First, the song underscores the widespread appeal of the velocipede

FIGURE I.5. Paul Hadol, "The Velocipedes," *La Chanson illustrée*, May 2, 1869.

and its ability to cross lines of class and gender. Next, the lyrics tie the velocipede to the machine age and the discourse of progress and utility, while at the same time humorously and anachronistically linking it with France's illustrious history. The lyrics also unabashedly reflect deep-seated sexism, equating female velocipedists with prostitutes—the tight-fitting and bright garb of the woman on the paper's cover attracts the lecherous gaze of a monocle-wearing bourgeois man beside her. Finally, Flan's verses conflate the velocipede with horses, a topic to which we will return in chapter 3.

Hadol's cover image for *La Chanson illustrée* additionally highlights the velocipede's role in pushing sartorial boundaries for women. Riding required freedom of movement that in turn mandated a new kind of wardrobe. Women could wear tights, short leggings, or even pants, though always within certain limits and not without criticism—a short chapter titled "Fashion and Velocipedes" in the *Manual of the Velocipede* provides recommendations for female fashion, but the man who authored it remains prescriptive in his advice and insists that a woman's "leg must be visible."[32]

An 1868 illustration depicting the first women's velocipede race showcases the new fashion afforded the riders (figure I.6). With their short flowing skirts, bloomers, tights, and caps, the riders challenge conventional fashion expectations. Compared to the dress of female spectators on the left of the illustration, the velocipedists enjoy a clear advantage when it comes to freedom of movement and comfort. But the image also conveys a sense of objectification: with fabric flying behind their shoulders, the riders look statuesque, like the famous Winged Victory (Nike of Samothrace) on display in the Louvre. And as the most prominent spectators in the front right of the image are all men, the male gaze further objectifies the pedaling women. This focus on the female body was not limited to aesthetics, however: exercise on a velocipede could help a woman maintain fitness, strength, flexibility, and good health, "so that she can acquire all the energy that she will need someday to conceive and give birth without danger" . . . and without her becoming muscle-bound like a female Hercules.[33]

Nevertheless, perhaps more than any invention, the velocipede promised new opportunities and freedom of travel for women, though this would later increase considerably with the advent of the modern bicycle.[34] The chapter "Velocipede Races" in the *Manual of the Velocipede* depicts a women's race and suggests that the velocipede inspired women of every class to seek greater autonomy. Other chapters describe

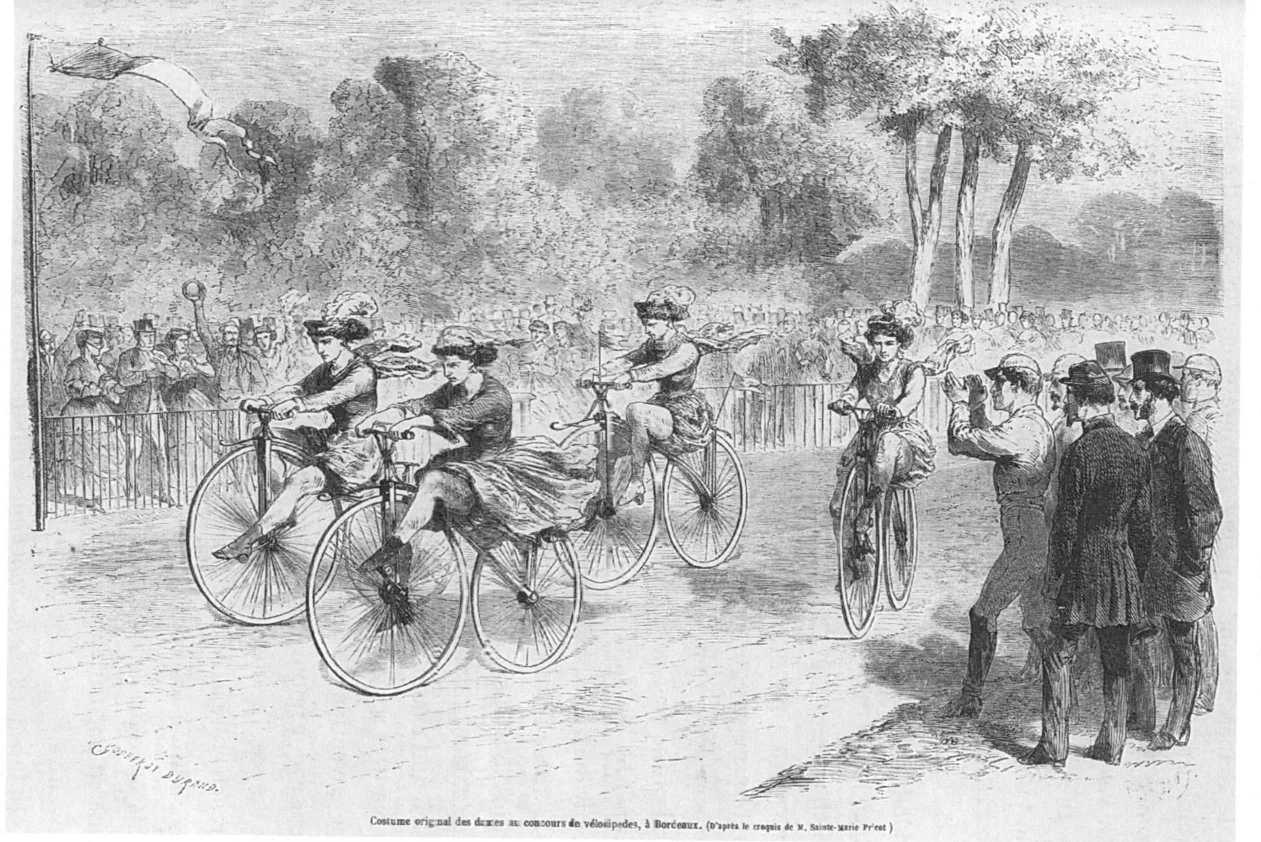

Costume original des dames au concours de vélocipedes, à Bordeaux. (D'après le croquis de M. Sainte-Marie

FIGURE I.6. Godefroy Durand, "Original women's attire at the Bordeaux velocipede race," *Le Monde illustré*, November 21, 1868.

female characters freely riding through Parisian boulevards without male chaperones and dressing transgressively (in 1892, the statute prohibiting women from wearing pants was officially amended for women on bicycles, but they had already begun flaunting the ban to ride velocipedes in the late 1860s).[35]

The impact of the velocipede even extended to the stage, one of the most important cultural spaces of the period. The opéra bouffe *Le Petit Poucet* (Laurent de Rillé, 1868) and the musical revue *Le Mot de la fin* (Clairville and Paul Siraudin, 1869) both featured velocipedes. In *Le Mot de la fin*, a velocipedist artfully and acrobatically rides around the stage; the script indicates that the velocipede is ridden by "the young Michaux"—one of Pierre Michaux's sons.[36] A review of *Le Petit Poucet* notes that a highlight of the performance is an escapade involving two amorous ogres on velocipedes. The review concludes by noting, "[The velocipede] is an instrument that will certainly figure prominently in end-of-the-year performances."[37] And in July 1868, a theater journal edited by the author Alexandre Dumas announced, "If we are correctly informed, the Theater of the Imperial Prince will present a revue with three or four tableaux [decors or settings] entitled *Les Vélocipèdes* from November to the end of December. The title . . . is very promising."[38] Henri Blondeau's operetta *Dagobert and His Velocipede*, translated and examined in chapter 2, was crafted around the velocipede. It features songs, puns, and quips about the velocipede and situates the invention of this nineteenth-century machine in the seventh century—a humorous way to embed the velocipede in the long history of France. The operetta also links the velocipede with other markers of French cultural capital: champagne, Gallic humor, and royal history. In addition to providing a window into a uniquely French genre, this operetta demonstrates the extent to which the velocipede had entered popular culture and how it embedded itself into the national narrative.

In June 1868, *Le Journal amusant* humorously described the impact of the velocipede on the French language itself:

> I velocipede
> You velocipede
> He velocipedes
> We velocipede
> You all velocipede
> They velocipede
>
> What I love most about our modern engineering inventions are their immediate consequences on the French language.
>
> Each new invention enriches the dictionary with a barbaric verb, a cacophonic noun, an adjective that resists honest rules of pronunciation.
>
> Take the velocipede, for example.
>
> Even though it was born just recently, it has already filled its quota: the verb *to velocipede*, plus the words *velocipedoman* and *velocipedomania*.[39]

The article goes on to describe people conjugating the verb *vélocipéder* in the present, future, and past tenses and even in the subjunctive mood.

The mania for the velocipede made its way to the upper spheres of French politics. When Émile Ollivier, the prime minister for Emperor Napoleon III, trailed in his election bid for a legislative seat in Paris to François-Désiré Bancel in 1869, the cover of the satirical weekly *L'Eclipse* featured Ollivier being bested by his Republican rival in a velocipede race (figure I.7). The note on the ground under Ollivier's wobbly machine reads "P.P.C.," meaning *pour prendre congé* or "going on leave." The article accompanying the cover illustration explains that Ollivier was losing because he had been willing to ride any velocipede, while his opponent rode a velocipede made by the "Compagnie Parisienne" (the descendant of Michaux's original company): "Look at Bancel. He will easily win. Thanks to his excellent mount (made of cast iron), he will keep his lead and his promises!" The article's author goes

Deuxième année. — N° 69 Un numéro : **10** centimes 16 Mai 1869.

RÉDACTEUR EN CHEF
F. POLO

ABONNEMENTS
PARIS
Un an.......... 5 fr. »
Six mois....... 3 »
Trois mois..... 1 50

Rue du Croissant, 16.

L'ÉCLIPSE

JOURNAL HEBDOMADAIRE

DIRECTEUR
F. POLO

ABONNEMENTS
DÉPARTEMENTS
Un an.......... 6 fr. »
Six mois....... 3 50
Trois mois..... 2 »

Rue du Croissant, 16.

COURSE ÉLECTORALE, PAR GILL.

ÉMILE OLLIVIER. BANCEL.

Voulez-vous arriver *beau premier* au but?... Ayez soin de choisir un vélocipède de la Compagnie Parisienne (*Ancienne maison Michaux et Cie*,) et vous serez certain de remporter le prix sur tous vos autres concurrents.....

FIGURE I.7. André Gill, "Electoral Race," *L'Eclipse*, May 16, 1869.

on to note that Ollivier's inferior machine caused him to "expend his energy in useless effort."[40] This is a reference to the fact that Ollivier spent much of his time campaigning for a second seat outside of Paris while also working in his role as prime minister for the emperor. Significantly, the article was penned by "Le Cousin Jacques," one of the many pen names of Richard Lesclide, the author of the *Manual of the Velocipede* (presented and translated in chapter 3).

Three months later, *L'Eclipse* returned to the velocipede and its connection to the state when Léon Bienvenu again used his satirical pen to address rumors that the state was planning to impose a tax on velocipedes:

> They say that next year, velocipedes will be hit with a 50 francs tax as two-wheeled carriages.
>
> Some advice for the new taxpayers:
>
> Wait until the tax collector requests you to come to the payment office.
>
> Make your way there by riding your vehicle.
>
> On the way, take care to run into a police officer and make sure you both fall into the gutter.
>
> The officer will arrest you and accompany you to the tax collector, to whom you will confidently declare: "If you please, I ran it through a city official this morning, as this gentleman can attest."[41]

Though the tax does not seem to have materialized, Bienvenu here uses it as a pretext to point out how divisive the velocipede had become in Paris and to poke fun at both velocipedists and Parisian bureaucrats.

In higher political spheres, Napoleon III's son was a fan of the new machine and rode his Michaux velocipede so frequently that he was known in Paris as "the little boy who pedaled everywhere."[42] Beginning in 1868, he was widely depicted in the press as a velocipedist (figure I.8), and by 1870 satirical journalists began referring to him simply as "Velocipede IV."[43] The illustrator Paul Hadol, who drew the velocipedists for the cover of *La Chanson illustrée* (figure I.5), later

COMPIÈGNE. — Récréation du Prince Impérial. (D'après le croquis de M. Moullin.)

FIGURE I.8. Louis Moullin, "Compiègne. Recreation of the Imperial Prince," *Le Monde illustré*, November 28, 1868.

imagined the imperial prince perched on a velocipede. In figure I.9, Hadol depicts the emperor's son as a canary—showy but useless; the French word for canary, *serin*, is slang for simpleton.

Finally, the *Manual of the Velocipede*, published in 1869, sets out to blend "the useful and the sweet," "instruction and entertainment," by anthologizing riding guidance alongside history, short theatrical sketches, purchasing instructions, and humorous stories.[44] Compiled and largely authored by Richard Lesclide—an indefatigable promoter of the velocipede—under the pseudonym Le Grand Jacques, it is the definitive work that both invents and cements the velocipede's place in French culture. Sociologist Jean-Marie Brohm has concluded that sport is an act of labor that imagines, in a capitalist society, the body "as a machine with the job of producing the maximum work and energy."[45] The *Manual of the Velocipede*, while it does acknowledge that the velocipede is an efficient means of travel, ultimately challenges Brohm's argument and suggests that sport, particularly sport on two

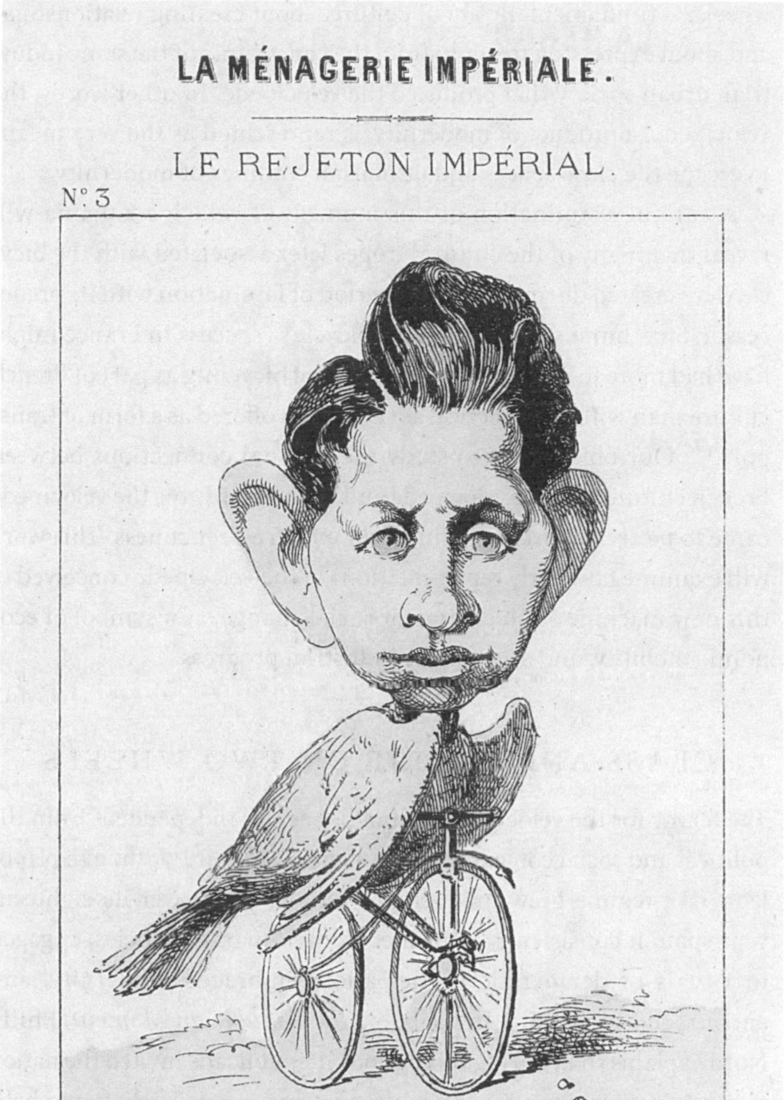

LA MÉNAGERIE IMPÉRIALE.

LE REJETON IMPÉRIAL.

N° 3

Au Bureau des Annonces, 11, rue Taitbout.

LE SERIN (Parade - Inutilité)

FIGURE I.9. Paul Hadol, "Imperial Offspring," *Collection de caricatures et de charges pour servir à l'histoire de la guerre et de la révolution de 1870–1871.*

wheels, is fundamentally about culture, about creating relationships, and about expressing freedom from the constraints of the same industrial, urban society that produced the velocipede. In other words, the velocipede, a product of modernity, is represented as the very means to escape the close spaces and utilitarian confines of modernity.

A careful examination of this moment of velocipedomania will reveal that many of the cultural tropes later associated with the bicycle were created during this short period of fascination with its predecessor. Smethurst suggests that the bicycle's "success in France might have had more to do with the perception of bicycling as part of French culture than with any practical advantages it offered as a form of transport."[46] Our objective is to study the original connections between French culture and the velocipede and to elucidate *how* the velocipede came to be seen as an embodiment of modern Frenchness. This work will examine how early representations of the velocipede conceived of this new machine as an engine for social change, as a symbol of economic mobility, and as a tool for industrial progress.

CLASS AND EMPIRE ON TWO WHEELS

The fervor for the velocipede cannot be seen as independent from the political and social context of the late Second Empire. Although Napoleon III's regime grew increasingly authoritarian over its eighteen-year span, it consistently implemented free-market policies, engaged in rituals of democratic participation, embraced modernity, and encouraged sociability. In his book *The Republican Moment*, Philip Nord explains that, during this period, "Republicans invited the nation to participate in a range of activities that encouraged beliefs and habits supportive of a democratic public life. The idea was to shape a particular kind of citizen: a conscientious human being who revered the *philosophes* and the revolutionaries of 1789, who valued liberty, laicity, and the riches afforded by literacy and a vital associational life."[47] These progressive mentalities of the late 1860s led to several important developments: an openness to modernization that allowed for new ideas in bureaucratic circles and new uses of technology, as we will see in the *Note on Monsieur Michaux's Velocipede*; the deregulation of theaters in 1864, which enabled troupes to explore different genres and react more quickly to contemporary trends, as we will discuss in the introduction to *Dagobert and His Velocipede*; and the rise of feminist organizations that pushed for rethinking of gender roles and female participation in public life, a movement that, as we will explain, resonates throughout the *Manual of the Velocipede*. It is important to underscore that the creation of new associations throughout France, facilitated by the slow relaxation of rules surrounding their official recognition and the legal authorization to create professional associations in 1868, created fertile terrain for the creation and proliferation of velocipede clubs.[48] These local velocipede associations elected officers, wrote bylaws that were approved by local authorities, and organized rides, races, and social events, contributing to the type of solidarity and participation that would be required in the soon-to-be-formed Third Republic. Alan R. H. Baker, in his 2017 study on French musical societies and sports clubs, notes that as early as 1868 velocipede clubs had already been formed in Paris, Castres, Valence, and Carpentras: "In supporting the [Carpentras] club's request for authorization, the sub-prefect claimed to the prefect that cycling was becoming *à la mode* throughout the whole of France and that already many people in Carpentras were taking it up."[49] In sum, the rise of the velocipede should be seen in parallel with the élan for liberty and sociability that characterized late Second Empire France. The velocipede both reflected and contributed to a contemporary interest in new technology and progress, to a bourgeoning feeling of independence, and to a sense of middle-class national solidarity.

Given the rise of an administrative and largely sedentary middle class in Second Empire France, people sought new ways to maintain physical fitness in urban settings. Gymnasiums began to spring up in Paris in the mid-1860s. Members could learn weight training, fencing, boxing, horseback riding, and, eventually, how to ride a velocipede. Eugène Paz, founder of a large gymnasium in Paris in 1865, adopted velocipedes as part of his gymnastic training as early as 1868, as is corroborated in the *Note on Monsieur Michaux's Velocipede.* Gymnasiums and the velocipede promised exercise, improved blood flow, and better health to city-dwellers who spent much of their day hunched over desks in stuffy offices. The connection between those early gymnastics clubs and the velocipede is evident in at least one key chapter in the *Manual of the Velocipede*, "On the Topic of Exercise," penned by Eugène Paz himself.[50]

ABOUT THIS BOOK

In chapter 1, we translate *Note on Monsieur Michaux's Velocipede*, one of the first sustained arguments to articulate how this new invention could be put to use in the service of the French empire and its economy. Chapter 2 examines the velocipede on stage and features a translation of the operetta *Dagobert and His Velocipede*, a work that demonstrates the extent to which the velocipede had permeated popular culture and how it could be used to parody both mythical French kings and France's contemporary cultural moment. In chapter 3, we present and translate the work that perhaps best epitomizes velocipedomania, *Manual of the Velocipede*, situating it in the literary, cultural, and political history of France. Finally, in chapter 4, we translate the earliest known poems about the velocipede—two originally published in French and one in Latin, all three at the height of the velocipede craze—and look at poetic representations of this new machine. Throughout this book, we attempt to show how early proponents of the velocipede argued for its usefulness and its beauty and how they embedded it in French culture in an attempt to give it staying power—an effort that would come to full fruition a generation later.

CHAPTER ONE

THE UTILITARIAN VELOCIPEDE

NOTE ON MONSIEUR MICHAUX'S VELOCIPEDE

The anonymous author of the *Note on Monsieur Michaux's Velocipede* identifies himself simply as an *amateur*: that is, a fan or an enthusiast. In a March 1869 article in the weekly periodical *Le Monde illustré*, a certain Monsieur de la Rue acknowledges having recently published a work on the velocipede. Given this admission and the identical writing styles between the article in *Le Monde illustré* and the *Note*, we can reasonably assume that de la Rue is the *Note*'s author. What's more, the article in *Le Monde illustré* is about de la Rue's invention, a nautical velocipede, featured on the periodical's cover (figure 1.1). De la Rue suggests it could be used for travel, sport, and lifesaving efforts and dubs it "the *insubmersible*."[1]

In the early 1870s, this same Monsieur de la Rue was named subprefect in St. Malo. In a profile for the newspaper *Le Figaro*, journalist Louis de Coulanges describes him in these terms:

> The Baron de la Rue [is the] inventor of the nautical velocipede—the day he invented it was the best day of his life. The illustrated press depicted him seated gravely on a nautical velocipede, breaking waves toward a better destiny. At the time, he was an employee at the Ministry of the Navy in the enlistment office; but he was seen as a mediocre employee. Velocipedomania filled all his time, paralyzed all his

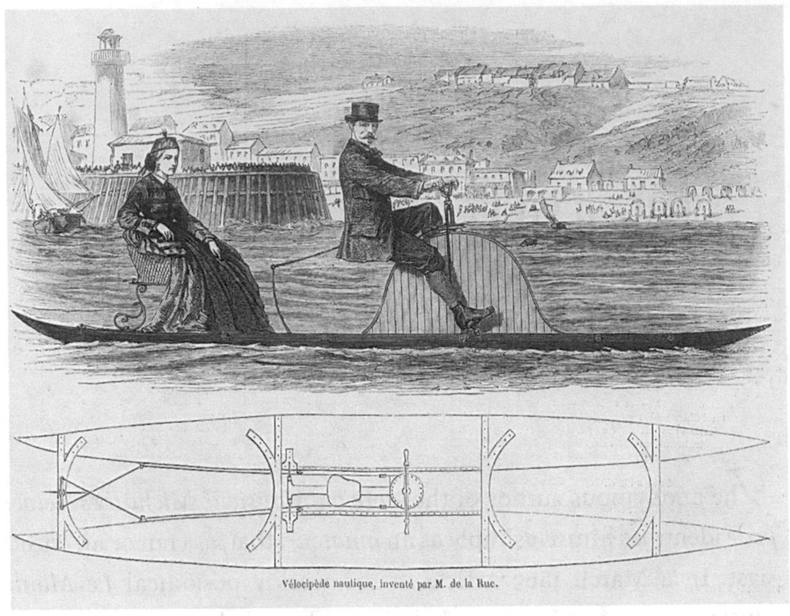

FIGURE 1.1. "Nautical velocipede, invented by M. de la Rue," *Le Monde illustré*, March 27, 1869.

> mental capacities. Each morning he could be seen arriving at the ministry on his two-wheeled horse, sweating and panting as if the fate of France were in the balance; each evening he went straight to the riding school of the famous Michaud [*sic*], the Gambetta of the velocipede.[2] In the office, he spoke only about velocipedes, thought only about velocipedes, breathed only velocipedes. When he would blow his nose, he did it velocipedically. . . . It appears he hasn't changed and that he spends his time navigating the port on his famous nautical velocipede (patent pending). In his neighborhood, they call him the subprefect of velocipedes.[3]

This less-than-flattering portrait confirms that Baron de la Rue is our author: he worked at the Ministry of the Navy, a detail acknowledged by the author of the *Note*, and he was indeed an enthusiast, someone who lived and breathed the velocipede.[4] The article also suggests that de la Rue may have published his treatise anonymously in order to avoid criticism for spending too much time on his velocipede and not enough time at his day job—a sensitivity every committed cyclist can sympathize with. The passage additionally presents a case study of someone suffering from the disease of velocipedomania: he is obsessed by velocipedes and can't think clearly because of them. Indeed, he has such a bad case that he must regularly blow his nose *velocipedically*.

In the *Note*, de la Rue acknowledges his passion for the velocipede and mentions his many excursions around Paris and longer rides in provincial settings—he is clearly someone who spent a lot of time in the saddle. His objective in writing, though, is primarily to underscore the velocipede's practical uses and to enhance administrative efficiency. He mentions the exhilaration of riding in passing but emphasizes how the velocipede can be used to improve the productivity of France's workers and facilitate swifter communication throughout the country. He is, in short, merging his passion and his work in an attempt to justify his mania.

Recollecting works published during the Old Regime that were dedicated to the King, the *Note* begins with an homage to the emperor, Napoleon III. De la Rue cites imperial decrees designed to create national transportation networks and adds that velocipedes will help unify the country and speed communication. The author of the *Note* is a model of the type of bureaucrat that came to prominence in the late Second Empire: an educated man with a fair amount of autonomy seeking to improve the French state through the implementation of new technology, primarily for the benefit of other affluent men of a certain position like himself. His belief in progress and improvement also points to the optimism and—in hindsight, given the Franco-Prussian War of 1870—the hubris that pervaded the upper-middle class in late 1860s France.

The untitled image on the cover of the *Note* (figure 1.2) underscores the work's utilitarian aspirations.[5] The illustrator chose men

NOTE

SUR

LE VÉLOCIPÈDE

A PÉDALES ET A FREIN

DE M. MICHAUX

PAR UN AMATEUR

PARIS

IMPRIMERIE DE AD. LAINÉ ET J. HAVARD

Rue des Saints-Pères, 19

1868

FIGURE 1.2. Cover of *Note sur le vélocipède* (*Note on Monsieur Michaux's Velocipede*), 1868.

representing three major groups of Second Empire France: in the foreground, an employee of the state whose uniform and demeanor signal the gravity of his profession; on the right, a well-dressed member of the bourgeoisie, still in his top hat and vest, who appears to be leaving the city to enjoy the space and fresh air of the country; and, on the left, a man whose gear and rifle evoke rural France. In a paragraph in the heart of the *Note*, the author confirms that the illustration depicts a cavalryman, an enthusiast, and a hunter. The author regrets not having asked the illustrator to instead depict an infantryman, a postman, and a telegraph worker, all state employees performing their work on foot who would be more productive with the help of velocipedes. Different characters on the cover, he remarks, could have even better underscored the transformative potential of the velocipede.

Throughout the text, the author provides examples of potential uses of the velocipede to improve productivity, communication, and the standard of living: mail can be delivered more expeditiously, emergency workers can be quickly gathered without the need to saddle a horse, and state employees can communicate with their correspondents more frequently and more efficiently. De la Rue effectively stakes out the practical applications of the velocipede, forecasting the many services—delivery, commuting, policing—that the bicycle still provides today.

It should be pointed out that de la Rue specifies in the work's title that he is describing "Monsieur Michaux's velocipede." As Coulanges mentions in the article cited earlier, de la Rue clearly spent a great deal of time at the Michaux riding school, and identifying Michaux in the title further cements the mythology of Michaux as the inventor of this new machine. But de la Rue also specifies that it is Michaux's velocipede "with pedals and brakes" to differentiate it from velocipedes without them. As mentioned in the introduction, the term "velocipede" had been in use in France to describe the draisine (a two-wheeled vehicle *without* pedals or brakes) and its variants. Once into the main body of his text, de la Rue drops the qualifiers for the velocipede, but then has to choose another term for Drais's running machine. Surprisingly, he opts for the term "celerifere," a term we contextualize in our endnotes.

This short defense of the velocipede was published in early 1868. Part of the text was also published in novelist Alexandre Dumas's newspaper, *Le Dartagnan*, in mid-May of the same year.[6] The syntax is frequently overwrought and the narration lacks the subtlety of the *Manual of the Velocipede*. In our translation, we have largely opted for legibility rather than mirroring the sometimes tortuous sentence constructions of the original. The section breaks and italicized phrases in what follows are duplicated from the original. Despite its shortcomings, the *Note* provides an important perspective into how the velocipede's earliest promoters angled to spread its use beyond the upper class and to advocate for its utility. This work's influence is apparent in other publications, most notably in the *Manual of the Velocipede*. We have left footnotes from the original as footnotes in the translation (here and throughout this volume); our explanatory notes appear as endnotes.

NOTE ON MONSIEUR MICHAUX'S VELOCIPEDE WITH PEDALS AND BRAKES

By an Enthusiast
1868

FOREWORD

One of the most intriguing economic principles, without a doubt, holds that improved communication leads to flourishing commercial activity and to a better quality of life for the entire populace.

The impetus behind the promise of improved communication networks comes from the Emperor [Napoleon III] who, via an 1861 decree, gave us national throughways and, in 1867, pledged the prompt construction of both regional roadways and new canals.

But that addresses only half the problem; what is most desirable is a complete solution. It is not enough to cover France with roadways; it wouldn't even suffice to establish regional railways. In my opinion, we need to encourage every possible means of locomotion currently available—while waiting for something better—to fully leverage the new roads. For example:

For transporting merchandise and travelers by land: the road locomotive—as trials have shown—would stand in for trains in areas where railways cannot be built.[7]

For transporting merchandise by water: towing lines along the riverways—in this respect the Seine is a particularly powerful example, and the popularity of *bateaux-mouches* proves their usefulness for travelers.

The increased use of these powerful implements will catch hold by itself, and in the near future, I like to think.

In a similar vein, there is also a new actor that scientists should perfect so that its influence not be limited by *human power** and whose propagation, even in its current state, demands our attention due to its usefulness: *the velocipede with pedals and brakes.*

In this note I plan to focus on this ingenious machine and hope to produce concrete results—if I am fortunate enough to make my deep convictions and extensive experience compelling.

I plan both to write in a contemporary vein—since the velocipede is in fashion—and to share the passion that motivates me.

And foremost, I hope to highlight all the velocipede's advantages instead of limiting myself, as others have done, to the pleasure it provides.

SECTION ONE

Definition, Origin, and Mechanics

The celerifere—responsive and quick, a sort of wooden horse placed on two wheels on which one balances while pushing forward by foot (*Bescherelle Sr.'s Dictionary*)—is not new.[8]

After being used some twenty years ago or more, it was abandoned because its flaws outweighed its benefits. However, we should be grateful to its inventor; we are indebted to him for an idea that has been revived today, though in very different conditions, as the velocipede; we will study the connection between the ancestor and its descendant.

Handling the celerifere was a bit like ice skating: tiring and even dangerous† for the rider, disastrous for shoes that obviously lacked skate blades and therefore didn't last long.

Handling a velocipede requires *balance*, *steering*, and *stopping*; these are all handled via a rod controlled by the rider's hands that, thanks to its simple but ingenious impact on the machine's core—the large wheel—acts as a *pendulum*, a *rudder*, and a *brake*.[9]

* In other words, the vehicle should become self-propelled. I am already convinced that the simple addition of blades could be used to propel the podoscaph (nautical sport).
† Wrong moves caused a tremendous amount of exertion.

Locomotion is accomplished by activating the pedals affixed to the hub of the *same wheel*; the propulsive power increases when the pedaling motion is smoother.*

The celerifere was difficult to steer, and it could not be controlled while on a descent; the velocipede responds immediately and can be stopped even on the steepest slope.

The name of the celerifere—meaning prompt, rapid—was mostly wishful thinking, while the velocipede deserves its name.†

In a word, the celerifere was the rough sketch of a good idea; the velocipede is the most thorough execution of that idea.

If any discovery can fill someone with a mixture of *shock and fear* at first and, as I boldly claim, with *admiration* later, it is certainly the velocipede with pedals and brakes.

I have often seen people in Paris and in the countryside—and every practitioner would give similar testimonies—who exclaim when seeing me on this new steed, "Ah! Monsieur, I would never want to try that."

But at the same time—with the exception of a few coachmen—everyone expresses admiration when they see the velocipede's *freedom* on the most congested roadways,[‡,10] on the busiest sidewalks,[§,11] and on the most serpentine, steep hills.[¶,12]

* Consequently, children are the most adept velocipedists.

† Two enthusiasts maintained an average of twenty-five leagues over the course of six consecutive days. Three others, *on a tour of France*, after training progressively, challenged each other: the first two stopped after thirty-five leagues; the third, in twenty-three hours straight, covered 199 kilometers.

‡ For example, I travel from the Bastille to the Ministry of the Navy on the rue de Rivoli every day in fifteen minutes.

§ I have gone down the sidewalk of the boulevard Sébastopol-Strasbourg several times without running into anyone.

¶ This is its ultimate triumph. The day before yesterday, descending the hill in Noisy, given my dizzying speed, roadworkers feared I would crash. A simple tap of the brakes reassured them; after slowing down I set off again to demonstrate my confidence and the compliance of my mount.

People's admiration is completely justified, and riding, rather than diminishing one's admiration, increases it by leading one to accomplish new feats each day, like the exploits outlined below that the inventor's sons reproduce for all-comers in front of their establishment on the avenue Montaigne.

The fear evoked is not the same* since, at first glance, it seems reckless, almost irrational, to risk getting on a mechanical horse whose two wheels, in the same plane, have a width of only two centimeters in contact with the ground; but this is *easily* overcome, thanks to the pendulum in one's hands that possesses a force as great as its suppleness, and *assuredly* overcome, since this mechanical horse has never said no to anyone.

As for the *potency* of the equilibrium on a velocipede,† let me report that Michaux's children have ridden with absolute confidence on the wall along the Seine and down the stairs at Trocadero, etc., and that several enthusiasts, without trying anything quite as dangerous, or, should I say, as useless, overcome equally challenging difficulties on a daily basis.‡

Based on what I have written above, I believe I can already allow myself to conclude that Michaux's velocipede is incredibly practical and, though its predecessor's popularity was short-lived,§ it has real staying power.

SECTION TWO

Practical Applications

The applications are as hygienic as they are pleasant, as useful as they are numerous.

* This fear is exaggerated by a number of beginners who want to be applauded for overcoming something difficult.

† When one acquires a sense of balance and counterbalance, it is never lost, in the same manner one never forgets how to swim.

‡ For example, while riding at full speed, they lie down or stand up on the saddle or let go of the handlebars and swing their legs over them in order to ride sidesaddle, etc.

§ As I hope will be the lifespan of the many counterfeits I see in the windows of toy stores that frighten me for the safety of their future riders on public roads.

FIGURE 1.3. Victor Rose, "The Great Gymnasium," c. 1866.

First and foremost, the velocipede is hygienic: the way it is propelled explains and fully justifies its popularity at the big gymnasiums of Triat and Paz—and undoubtedly at other gyms—where it is considered one of the best instruments for developing the muscular system (see figure 1.3).[13] The words of one of the most talented doctors in Paris corroborate this practice: "Each time I see this instrument, it makes me dream: I want all my patients with gout, with rheumatism, with neuropathy to have one; I hope everyone who suffers with weakness will use one. If I were a father, I would give one to my children."*

As for *enjoyment*, like horseback riding,† like hunting, etc., like everything that is most inspiring, the velocipede excites people.

I could wax poetic about the pleasures derived from the velocipede during my morning rides around Paris when I discover ravishing

* It seems very desirable for this type of exercise to be taken up by schools and by families.

† Except the velocipede is more docile than the best-trained horse: it isn't irritable, never bolts, and can be loaned to anyone. It is always shod and saddled, doesn't need food, grooming, or veterinary care. For speed and endurance, it can rival most of its flesh-and-bone colleagues.

places that I would have never noticed without it. I could relate my longer excursions along the coast, in Burgundy, and under the beautiful sky of southern France. I could endeavor to describe the charm of daily rides on this mount and how, at will, one can accelerate, slow down, or stop, depending on whether one wants to leave a barren grassland or a scorched plain behind, admire a beautiful landscape, rest or dream in the shade of ancient trees, spend time in a picturesque countryside, admire artistic curiosities, etc. Doing so would unnecessarily exceed the scope of my argument since, when it comes to enjoyment, the case of the velocipede has already won the argument in France and abroad.[*,14]

Among the velocipede's most fervent disciples are princes, dukes, marquises, etc.; athletes, senior civil servants, a member of the academy; even the most distinguished have not scorned it.

In its honor, I'm told a club will soon be organized and races established;[†] we can therefore trust that its popularity will continue. And I am sure that a model in aluminum bronze will be developed, a model that its predecessor inspired me to imagine: it will be elegant, resembling polished gold; it will provide steel-like stiffness and more elasticity; and most of all, it will make maintenance much easier.

But will the velocipede remain simply an instrument for excursions, sport, or luxury? I hope not. *It also aims to be an auxiliary to work.* It must trickle down to every social class in order to yield real benefits. Just as it found sponsors in the *high life* for the pleasure it provided, it likewise has intelligent applications among the industrial and working classes. Public administrators wisely encourage its use, *a neces-*

* Today one can find velocipedes all over Europe. It is even known overseas. One of my good friends has been using one for two months in Cochinchina, shocking the local residents. It has already caught on among the officers of that beautiful colony.

† There have already been some in Cannes; others are planned in Billancourt.

sity for every new invention, and I would be happy to serve as its modest promoter.

The charming sketch made by a dear friend who lent his talents to this work's cover presents three types of velocipedists: an *enthusiast*, a *hunter*, and a *cavalryman*.

We can see that the two former riders, having spent a long time in the saddle, are in their element. This is not the case with the cavalryman, who appears ill at ease on the velocipede: either he did not want to be unfaithful to his usual mount by riding its rival or he is not yet familiar with this new vehicle.

But since I didn't express myself as clearly as I wanted, I ask my friend to please forgive me.

I was hoping that the image would have a foot soldier, not a cavalryman, since a velocipede would be redundant in the cavalry, but it would be precious to a company of infantrymen detached to an outpost, a fort, etc. They could use it to communicate with military headquarters, for reconnaissance, etc.*

I also noticed, but too late, that instead of depicting a tourist and a disciple of St. Hubert,† it would have been more useful for my argument to have represented a rural postman and a telegraph operator on velocipedes.[15]

The velocipede could transform rural deliveries, a service becoming both more important and more arduous. Quicker travel would lead to a triple benefit: for the *public*, since the work of an entire day could be done in several hours; for the *civil service*, since the speed of communication would multiply its efficiency; and for the material

* I leave to experts the task of identifying the velocipede's usefulness in the service of the army.

† Depending on the length of their legs and, *to some extent*, on the terrain, hunters will gain muscular strength and will be able to travel four or five leagues an hour while carrying a twelve-gauge shotgun, a pack, and a full game bag on the return journey—a valuable guarantee if not for the privileged sportsman who hunts on the outskirts of his property then, at least, for the inhabitant of a barren rural area (Brittany, for example), where a long trek is required before leaving the road for hunting ranges.

well-being of the *agent* himself, since he could use the time saved on deliveries to do other work and augment his salary, assuming his obligations were calculated taking into account his new locomotive ease.

I predict some objections to which I would like to respond.

Not all roads are passable on a velocipede. Today there are very few that aren't passable;* soon they will all be.[†,16] Moreover, even if the planned improvements were carried out only in three of ten districts, their importance could not be overstated.

Postmen often take paths that remain inaccessible to velocipedes. But once they have this new means of locomotion, they will no longer need to take these shortcuts. Even if they have to travel a third farther, it will still be quicker to ride a velocipede.

If someone asked me what would happen to a velocipede were it to be momentarily left for some reason, I would reply that it can be chained up, locked, put away, leaned against a wall, or laid in a ditch where it awaits the return of its rider with no impatience and with no fear of being taken.

I have already addressed the perceived difficulty of riding a velocipede, maintaining that one can easily learn to ride it at any age and that it is, without question, less tiring than walking or riding a horse.

Finally, one may say that ascents stop velocipedes. To climb a slope over four centimeters per meter, it is true that one must dismount and push it by hand. But even then, it remains a help rather than a hindrance since—at the risk of sounding paradoxical—I attest that in a climb over two kilometers long, from Villers[-sur-Mer] to Houlgate (in the Calvados region), I leaned on it as I walked up the hill and thus limited my fatigue.

I can therefore confidently bring this vehicle to the attention of the clever and committed administrator who told the legislature last year,

* A road must be full of potholes to stop a velocipedist. I have ridden on paved roads that lack the care provided by Parisian authorities, and on eroded or furrowed paths with icy ruts, without the slightest accident.

† On April 7 this year, *La Patrie* published an article reminding readers of the proposed law to enhance country roads.

"The government's trend has been to accelerate the delivery of correspondence and to improve conditions for postal agents. *Making history requires forward thinking.*"

I could also apply these words to the Director General of Telegraphs. Like his colleague at the Post Office, Monsieur the viscount de Vougy has already done a great deal for his agency.[17] For him, too, making history requires forward thinking.

The future is *present* in the form of the velocipede, which, if it is used for transporting dispatches, would assure new and desirable results for the public, for the administration, and for agents:

For the *public*, because people would no longer need to wait several hours for a telegram; if the receiving office is three or four leagues away, or farther, the telegram must be handed over to the postal service, necessitating more expense and a waste of time while it is delivered by a postman on foot.

For the *administration*, since the speed of electric communication would be complemented by the speed of the velocipede: telegrams could be multiplied by a factor of ten.

For the *agents*, even if their compensation were reduced, since they could complete three or four routes in a day compared to the one they complete now.

But these already decisive advantages represent only the beginning of the velocipede's utility.

The Aquatic Rescue Society, founded in 1865 under the leadership of a distinguished admiral, continually strives to improve its organization and its methods for carrying out its humanitarian work: it adds new lifeboats each day and now has life preservers along with mooring line arrows and cannons to rescue ships in distress.[18] It has another new technology at its disposal: the semaphore,* used to signal ships

* It is referred to as a veritable trumpet of the final judgment, since it can be used in storms, regardless of their fury.

and even prevent disaster. It should also have the velocipede in order to rapidly gather crews for the lifeboats, something that can be easily undertaken if the distressed ship is in or near a port, since in these moments ten brave men will unite to save one in danger. This becomes difficult, unfortunately and as is typical, when sailors have to be alerted in their cottages, which may be spread out far from the point of embarkment.

It would also be highly effective to employ the velocipede for surveillance of our *coasts, tributaries and rivers, forests, large estates*, etc. The velocipede would allow guards to quickly and spontaneously check areas that are seven or eight leagues from their residences, places they rarely visit today but that they would visit with pleasure once we have provided them with an enjoyable form of travel.

The velocipede would be incontestably useful in the case of fires—sadly, all too frequent—since it is always ready and could be far down the road, on the way to providing emergency relief, before any other speedy form of transportation could be ready to leave.

It would help with local travel. People already prefer it to their carriages, or at least, like me, they prefer it to the public coach for going to the train station and back. We will soon see people giving velocipedes to their servants so they can run errands that have required the use of a horse up until now.*

Once it has overcome prejudice, once it is formally granted citizenship, the velocipede will supplant or supplement the country doctor's horse; at that point, I would also like to give it to the priest who covers a second congregation far from his home parish.

* Factory workers are already using them.

Finally, if its price could be lowered—since the only downside is that it is out of reach of those it would help the most—I would confidently recommend it to agents of the Civil Engineering Administration, to surveyors, road workers, etc.—in a word, to any worker who needs to make long trips that cause fatigue and take time.*[19]

PRACTICAL ADVICE

After reviewing the diverse applications that could be made of Michaux's ingenious device, I could stop; however, moved by the desire to have his machine understood and used, and authorized by the results I've seen, I believe I must reproduce here, in its rough form, the advice I sent along with a velocipede to a friend who is currently in Cochinchina and whom I have already mentioned in this study:

> Your velocipede is on the way, and you will be overjoyed. Since you'll have neither *a teacher nor a training device* where you are, I insisted they send you two different metal strips for the saddle: one is low enough so that, should you lose balance, you will be able to catch yourself with your feet; the other, much less curved, is designed for an experienced rider.
>
> At the sight of the device, you will understand that everything is a question of balance and confidence. I won't say anything about the latter since you will gain it on your own: thanks to the *training strip*, you will always land on your feet.
>
> As for balance, it is easy enough: it's all in your hands, and that should initially be your *exclusive* focus.
>
> Despite the fact that I would love to see you, I won't be able to go all the way to Saigon to share lessons from my experience with you, but here is some advice.

* I'm told by the publisher of this Note that the typesetters for *Galignani's Messenger* use velocipedes to go to work.

1st: Run next to your mount as you push it to get used to its movements; do this for several minutes.

2nd: Push the mount on a slight downhill so that it can go by itself. Push off and straddle it, legs dangling, and focus on the handlebars, a true pendulum that will restore balance when moved as needed.

3rd: When your hands gain experience, you will have mastered balance; next put a foot on the pedal and try to *follow its movement without pushing*. Then find flat terrain and you will become the agent of your own locomotion. You will have completed your velocipedic education without a fall as I promised you; and, like all who know this enjoyable and hygienic instrument, you will soon become a fanatic.

A final word: at first you will use far more effort than necessary. This is always the case with poorly performed exercises. In the movement of your feet, mostly when you stop pedaling, there will be stiffness, and your speed will suffer as a consequence. But don't worry. Whatever your aptitude may be, practice will allow you to relax, and that will in turn lead to perfection.

That, in my opinion, is the charm of the velocipede. It isn't purely mechanical. It allows your intelligence, your courage, and your skill to shine.

CONCLUSION

How will this notice be received? I can only hope it will be viewed as independent and conscientious.

I'm happy to conclude as I began: that the velocipede with pedals and brakes, agent of lively and salutary pleasure, can also be used as locomotion for serious and *practical* advantages. It merits the attention of individuals with initiative and spontaneity, the attention and solicitude of the government, and even the attention and the distinguished patronage of the sovereign; I at least hope to rally the support and voices of my colleagues who also participate in this type of equestrian exercise.[20]

To those who see what I have written as an advertisement, I can give you the satisfaction of conceding that that was my goal. This concession is easy to make: I'm not expecting any bonus from Michaux, the Varenne racecourse, or others. I was inspired only by a conviction authorized by facts *guaranteed by my sources* and by my desire to help create *economic* and *philanthropic* progress, something much more significant to the public than to an honest and talented artisan—Monsieur Michaux—whose future is already assured.

CHAPTER TWO

THE VELOCIPEDE ON STAGE

DAGOBERT AND HIS VELOCIPEDE

The fall 1868 theater season saw a boom in the number of opéras bouffes performed in and around Paris. An 1864 decree had deregulated theaters, and opéra bouffe became the genre that most reliably filled seats in this newly competitive market.[1] These comic musical productions had fewer rules governing them than productions in state sponsored theaters and they tapped into a desire to see parodies of history, of well-known fairy tales, or of other operatic works. It could be argued that the defining characteristic of the opéra bouffe and its direct descendant the operetta can found in the way they satirize other musical works. Jacques Offenbach's 1858 *Orpheus in the Underworld*, for example, the work that effectively launched the opéra bouffe in Paris, parodies the myth of Orpheus and Eurydice while spoofing music from the classical repertoire.[2]

That fall, theater critics remarked that the rise in popularity of the opéra bouffe paralleled the rise of the other popular fad of 1868: the velocipede. Writing in *L'Indépendance dramatique* in September of that year, P. de Faulquemont summarized the theatrical moment in these terms:

> It has become clear that velocipedes are going to take over our theaters.

> At the Théâtre de l'Athénée, in *Le Petit Poucet* [Little Thumbling], [the actors] Léonce and mademoiselle Lasseny make their entrance on velocipedes and sing a duet.
>
> At the Théâtre de la Gaîté, in the play *Nos enfants* [Our Children], Monsieur Gaillard appears on a velocipede and apparently handles it with a high degree of skill.[3]
>
> Finally, at the Théâtre des Menus-Plaisirs, in *Croqueuses de pommes* [Apple Munchers], one can see no fewer than twelve velocipedes on stage at one time.[4]
>
> To give credit where credit is due, we must recognize that the idea originated with the authors of *Le Petit Poucet* whose play has been running since January.[5]

Three weeks later, theater critic A. Vautier quipped that the genre of the opéra bouffe should be forever paired with the velocipede since both seemed to be everywhere:

> The opéra bouffe and the velocipede will soon become two inseparable companions—one called to reform society, the other to reform literature.
>
> As a result of the free market, the cost of velocipedes will continue to drop. Everyone will be expected to have a velocipede. One won't leave home without a hat and a velocipede.
>
> The cook will go to the market on a velocipede. Low-level employees earning 12,000 francs will travel to their offices by velocipede. Law clerks will go by velocipede to the corner butcher to buy their slice of ham. . . .
>
> Consider the results:
>
> There will be no more omnibuses, no more carriages, no more horses. The horses will be slaughtered and meat from the butcher will be sold for half price.
>
> The omnibuses and cabs will be chopped up; the poor will be able to heat themselves for ten winters in a row with all that wood.

Where before one could find hay and oats, farmers will now plant cabbage, and they will travel to their fields on velocipedes. . . .

We have just invented the aquatic velocipede; we will soon see the aerial velocipede, the subterranean velocipede, etc.

For its part, the opéra bouffe will cause the same type of revolution in literature. Following the lead of the velocipede, the opéra bouffe will take over every theater. We hope to soon see the genre go from the Théâtre des Variétés to the Vaudeville. And from there to the Gymnase, then the Odéon, then the Théâtre National Français, then the Opéra Comique and then the Opéra. And then finally to the Théâtre Cluny, the last literary theater in Paris. . . .

And now, to the music!
Nothing is such a treasure
As traveling for pleasure.
And all you need
Is a velo—a loci—a cipede
a velocipede![6]

Both of these articles reference a scene in *Le Petit Poucet* and point to it as the first instance of the velocipede being integrated into the opéra bouffe. Even though the lyrics from this scene are actually misquoted by Vautier, their repetition points to the popularity of the velocipede song and the way it resonated with the theater-going public. A review in *Le Figaro*, published in October 1868, describes the scene as follows:

Madame Krockmachcru, the ogre's wife, is named Aglaé, a courtesan who also happens to be married. She has the morals, the language, the feet, and the appetite of a prostitute from the Mabille Dancehall. . . . She swindles a few rare coins from an ogre in the neighborhood named Rastaboul (pronounced *Rase-ta-boule* [Shave your dome] to add to the humor of the character). The most memorable stunt the

> two lovers pull is a carefree velocipede ride. Alas! To survive today, the dramatic arts are reduced to integrating whatever happens to be in fashion. . . . Soon, every play will be forced to use velocipedes in order to succeed.[7]

Despite his disgust with what he views as pandering to the audience by acquiescing to the fad of the velocipede, the critic nevertheless concedes that the scene and song with velocipedes are the most memorable of the entire production.

Here, then, is the "Duo des vélocipèdes" from *Le Petit Poucet*, with lyrics by Eugène Leterrier and Albert Vanloo and music by Laurent de Rillé:

> Rastaboul: Nothing brings so much pleasure
> Aglaé: Nothing brings so much pleasure
> Rastaboul: As traveling for leisure.
> Aglaé: As traveling for leisure.
> Rastaboul: And all you need
> Aglaé: And all you need
> Rastaboul: Is a velo
> Aglaé: a loci
> Rastaboul: a cipede.
> Aglaé and Rastaboul: One, two, a velocipede.
> Aglaé: We're free to go where we will
> Rastaboul: Even though we might take a spill.
> Aglaé: We straddle
> Rastaboul: The saddle,
> Aglaé and Rastaboul: We straddle the saddle.
> Aglaé: When you ride over rough terrain
> Rastaboul: Your hands and your back are in pain.
> Aglaé: We have scares.
> Rastaboul: But who cares?
> Aglaé: Let's repress the painful sting
> Rastaboul: Let's repress the painful sting

Aglaé: And together we'll both sing . . .

Rastaboul: And together we'll both sing . . .

Aglaé: In rhythm . . .

Rastaboul: I'll begin!

Aglaé (*spoken*): Excuse me! You always begin! Now it's my turn. (*Singing.*) Nothing brings so much pleasure

Rastaboul: Nothing brings so much pleasure

Aglaé: As traveling for leisure.

Rastaboul: As traveling for leisure.

Aglaé: And all you need

Rastaboul: And all you need

Aglaé: Is a velo

Rastaboul: a loci

Aglaé: a cipede.

Rastaboul and Aglaé: One, two, a velocipede.[8]

Despite the silliness of two ogres riding velocipedes onstage, this song points to two themes that go hand in hand with the velocipede through the late 1860s and beyond: almost limitless freedom of travel and physical pain. The monstrous couple evokes the pleasure of traveling wherever they desire on the new machines. They also refer to the ever-present danger of crashing, and to the back and hand pain that every velocipedist knew only too well. In fact, the line we have rendered above as "We have scares" is more literally translated as "I am dead!" The physical experience of riding a wooden and steel velocipede over cobbled or gravel roads was certainly not foreign to the librettists.

At the end of 1868, the velocipede made its way into a variety show put on by an amateur troupe at the Salle Molière: a showcase in five acts and seven tableaux (settings or decors) titled *Paris-Vélocipède*. The most extensive description of this end-of-year revue—in fact, the only one to mention its use of a velocipede—appeared in the *Théâtre-Journal* on January 10, 1869. Julien Deschamps notes that the authors, Gaston Marot and Charles Gobert, did well to select a title that reflected the moment: "It's the best title of the year-end revues. *Paris-Vélocipède*! It's current, it's big, it's the hobbyhorse of the day!!"[9] While much of Deschamps' review focuses on the costuming, he does mention several actors who impressed, including Monsieur Duval, who "velocipedes agreeably."[10] Other reviewers also mention the costuming and some of the young singers, though nothing else is made of actual velocipedes.[11] But the revue is significant because it serves as a reminder that the velocipede became the de facto symbol of Paris in 1868 and 1869.

In early 1869, another revue at the Menus-Plaisirs theater sponsored by the newspaper *Le Figaro* also featured a velocipede. *Le Théâtre illustré* notes that one of the Michaux sons performed on his velocipede and that "his dexterity agreeably surprised the spectators."[12] The same periodical published an illustration of Michaux's son dressed as a jockey, reclining on the velocipede with his hands off the handlebars and his feet off the pedals, surrounded by other performers from the revue (see figure 2.1). *La Comédie* calls Michaux's performance "a marvel."[13] And *Le Petit Marseillais* raves, "The authors of the *Figaro-Revue* have managed to incorporate a velocipede solo into their play that the young Michaux executes each night with remarkable brio to the great satisfaction of the crowd."[14] Michaux was busy that winter: later in January he also performed at the Théâtre des Variétés in another revue titled *Le Mot de la fin* (mentioned in the introduction).[15] *L'Indépendant français* praised the entire production, concluding with this appreciation of Michaux: "I forgot to mention the famous Michaux, Jr., the noted velocipedist; he zigzags with such grace, and elegantly traces curves across the stage! He has just one flaw, namely, he is too much like a will-o'-the-wisp. I find myself obligated to apply this line: *He just arrived and already he's gone!*"[16] Again, the presence of a velocipede in these year-end revues underscores the extent to which the new machine had entered the Parisian cultural zeitgeist of the time.

It was in this theatrical context that the Théâtre des Petits-Bouffes Saint-Antoine opened its doors under a new name and under new direction in the fall of 1868. Director Ernest Martin commissioned an operetta that capitalized on the invention that was all the rage in Paris. For this popular theater located just a block away from the place de la

FIGURE 2.1. "Figaro-Revue," *Le Théâtre illustré*, January 1869.

Bastille in Paris, *Dagobert and His Velocipede* became the hit of the fall season.[17] Critics raved, and the theater was packed. Writing the week of the premiere for his journal *La Comédie*, Paul Ferry exulted, "The new hall was overflowing. . . . There was enthusiasm in the audience and on stage. . . . *Dagobert and His Velocipede* is a full-fledged operetta, the hit of the evening, an act that should be performed all over the country."[18] Ferry praised Martin, who both directed and starred in the role of Dagobert, for singing with "brio and originality."[19] He also signaled the contribution of the up-and-coming composer Frédéric Demarquette, suggesting he could become "another Offenbach"—the most celebrated composer of French light opera.[20] A review in *Le Figaro* praised *Dagobert and His Velocipede* for its humor, for the way it incorporated the popular song "Le Bon Roi Dagobert" into the score, for the performance of Mademoiselle Ambroisie in the role of Lucette, and for the script's clever use of anachronisms.[21]

Indeed, much of the humor in this operetta stems from its exaggerated and abundant use of anachronisms. Although the play is set in the Middle Ages—in 638 to be precise—the language includes nineteenth-century slang and references to sewing machines, to champagne houses founded in the eighteenth century, and, most notably, to the titular and most modern of all vehicles, the velocipede. The velocipede is presented as having been invented by King Dagobert's minister, Eligius, in the seventh century. What's more, Dagobert has access to the *Dictionnaire Bouillet*, a popular nineteenth-century encyclopedia that included an entry on Dagobert himself. Whenever he wonders what to do, he looks at his *Bouillet* to see what he did, then he does it. And indeed, Dagobert does feature as an entry in the 1863 edition of Bouillet's encyclopedia:

> Dagobert I, son of Clotaire II, became king of Austrasia in 622 and added Neustria to his reign in 628 at the death of his father, and Aquitaine at the death of his brother, Caribert, in 632. He defeated the Saxons, the Gascons, and the Britons; but he tarnished the glory of his victories through his cruelty and his passion for women. He founded

the Saint-Denis Cathedral in 632 and was buried there in 638 at his death at age thirty-six. Dagobert helped the arts flourish, especially sculpture and goldsmithing. Saint Eligius, who had been a goldsmith, was his prime minister and friend.[22]

The velocipede is used to humorous effect in the operetta. On one hand, the velocipede remains a symbol of the ultramodern; it stands as the embodiment of 1860s French design, technology, and freedom. On the other, set in the seventh century, the velocipede can be seen as grounded in the far-distant past and connected to the ever-present French cultural values of independence, seduction, and revelry. In the 1869 *Manual of the Velocipede* (chapter 3), Le Grand Jacques similarly ties the velocipede to a distant past in an effort to weave it into the fabric of French history and to give it cultural currency.

Why did the playwright and lyricist choose Dagobert over other French kings like Clovis or Charlemagne? First and foremost, because a song from the revolutionary period popularized the image of "Good king Dagobert . . . who puts his knickers on backward" ("le roi Dagobert qui met sa culotte à l'envers").[23] Fortunately for Dagobert, according to the lyrics, his friend and prime minister, Saint Eligius, was there to help the king overcome this sartorial mishap. To this day, French children sing about the good king Dagobert, his knickers, and his friend Eligius. This song, a staple of the nineteenth century, had many verses and became a popular way for members of disempowered classes to playfully mock the nobility. Dagobert was a safe king to mock. He lacked the religious, political, and cultural significance of a Clovis or a Charlemagne; and since he was backward (or at least his knickers were), he could be read in contrast with these so-called great kings of France. In the French imaginary, Charlemagne is the serious founder of French monarchical, religious, and legal practices. Dagobert, like the song suggests, has it all upside down. He is the carnivalesque king, one known for flaunting piety in order to pursue his passions: cruelty, women, song, and drink! Given that the velocipede was often associated with Carnival, the pairing of Dagobert with the velocipede makes sense, particularly in 1868 when the monarchy is decidedly out of fashion and the velocipede's popularity is on the rise.

Dagobert and His Velocipede ran until at least the end of November 1868 and was a big enough success to justify publishing the score and the script the following year in order to make it available to other theater troupes.[24] The published edition of the operetta indicates that the musical score for piano with lyrics was available for purchase separately and that the original orchestral arrangement could also be provided for larger theaters.

While *Dagobert* is not part of the repertoire of famous opera houses around the world like works by Offenbach and Bizet, it is nevertheless a great example of the French operetta bouffe, incorporating many themes that define the genre (historical parody, anachronisms, drinking, variations on well-known musical motifs). Unlike the more popular *Le Petit Poucet*, *Dagobert* treats the velocipede as a central theme and it is important from beginning to end of the libretto. And unlike *Paris-Vélocipède* and the other revues in late 1868 and early 1869, *Dagobert* is more than just a series of sketches: it has a coherent narrative and musical themes that reflect the playful way the velocipede was represented in French culture at the time. This humble but funny operetta bouffe is an ideal example of the type of productions the velocipede inspired in the late 1860s and it functions as a window into the cultural uses of the velocipede in the most popular genre of the day.

The published script includes a note that details the special effects to be used in the production: "The velocipede on which Dagobert makes his entrance is a small cart with two spinning wheels attached above it. Between the two wheels that imitate those of a real velocipede is a small seat where Dagobert sits. The cart's casters are hidden by the upstage wall that is between fifteen and twenty centimeters high. When the character enters, he is pulled by a rope from the left to the right of the stage. As he goes offstage, he takes a real velocipede of the same color as the one on the cart and reenters pushing it by hand."[25] The velocipede is a key element of the narrative. It enters with Dagobert, it plays an important role in the story of Caribert (Dagobert's deceased brother), and it is central to the operetta's denouement (though we won't give that away here).

It is worth remembering that Napoleon III and his court were associated with the velocipede: in May 1868 velocipede races had been held at Saint-Cloud, the site of an imperial residence.[26] Though parodying the imperial family would have immediately attracted the attention of the censors, depicting a long-dead royal was innocent enough. In fact, in his review, Jules Prével remarks, "Under the reign of Emperors, censorship . . . allows jokes to be made about kings."[27] Dagobert, like Napoleon III, was criticized for his many affairs and his cruelty, but these alone would not have established any clear parallels between the seventh-century good king and the contemporary emperor of France. Nevertheless, the velocipede and its connection to paternity in the last scene of this operetta would have likely brought the imperial family to mind, however tangentially. Given that the emperor's son was synonymous with the velocipede, the final scene causes a rethinking of all the anachronisms that came before: What if this medieval farce is a thinly veiled send-up of the current resident of the Tuileries?[28] French history may be filled with despotic womanizers, but adding a velocipede to the mix narrows the field considerably.

In addition to the many anachronisms, the operetta's humor is also built around puns, many of them real groaners. The script is self-conscious of the exaggerated nature of these puns and uses them as a sort of marker for upper-class wit. The full import of the puns is revealed only in the operetta's final scene. We have done our best to adapt the puns into English, and in most cases we explain the originals in the notes. We have included italics for some of the puns: these italics are also in the original script and underscore some wordplay that might otherwise go unnoticed.

Dagobert and His Velocipede was an early success for its lyricist and playwright, Henri Blondeau (1841–1925), and is one of the only operettas he penned by himself. After writing song lyrics and some sketches, he began collaborating with Hector Monréal, with whom he would create over thirty operettas, and, famously, the lyrics for the song *Frou-Frou* in 1897. The lyrics for this song suggest that the velocipede remained one of Blondeau's inspirations:

A wife sometimes wears
Knickers in her own home.
This fact is observed, I believe,
Within the bonds of marriage.
But when she goes pedaling
In knickers, like a Zouave,
It seems more *grave*.
[. . .]
Wearing knickers, you might say,
Makes riding a bicycle more comfortable.
But I say that without *frou-frous*
A woman is not complete.
When you see her pull up her skirt,
Her petticoat puts you under a spell.
Her *frou-frou* is like the sound of wings
Fluttering by and caressing you.[29]

In their *chanson*, Blondeau and Monréal capitalize on the new fashions the velocipede introduced in the 1860s and nostalgically allude to the eroticism evoked by the velocipede and its concomitant fashion, outlined in detail in the *Manual of the Velocipede* by Le Grand Jacques.

In fact, it appears that Blondeau and Le Grand Jacques (Richard Lesclide) crossed paths frequently. Blondeau covered races and wrote detailed articles about them in Lesclide's newspaper, *Le Vélocipède illustré*, throughout 1869.[30] What's more, they were both judges for the Paris-Rouen velocipede race in November 1869.[31] Blondeau may have seen writing a libretto about the velocipede as both a way to get some laughs at the expense of a popular fad and as an attempt to further promote the new invention that he would advocate for in the press.

Finally, this is an operetta bouffe: *operetta*, meaning performers alternate between spoken dialogue and song, and *bouffe*, indicating that it has a comical subject matter. All the verse sections were sung—with only one short exception at the end of the play. We have attempted to respect the meter and the rhyme schemes of the original songs. The piano score with the French lyrics can be consulted via the Bibliothèque nationale's online collection, Gallica.[32] In addition, we have recorded the music and made it available online at https://velocipede.byu.edu.

DAGOBERT AND HIS VELOCIPEDE

Text by Henri Blondeau
Music by Frédéric Demarquette
Operetta Bouffe in One Act

Premiered at the Théâtre des Petits-Bouffes Saint-Antoine
September 1, 1868

CAST:	ROLE CREATED BY:
Dagobert, King of France, 40 years old	Ernest Martin
Eligius, Minister, 50 years old	Riga[33]
Goldbrick, Troubadour, 20 years old[34]	Deberg[35]
Weathervane, Bourgeois, 50 years old[36]	Briand
Sheriff	Schoux
Lucette, Weathervane's daughter, 16 years old	Ambroisie
Two Soldiers	
Two Clerics	

Costumes designed by Marcel and made by Landolph. Set design by Vaillant.

Period costumes with as much whimsy as desired.

The action is set in the Champagne region of France at the home of Weathervane around the year 638.

The stage is designed as a garden courtyard. A tree is at left; on the right a pavilion in the style of the period with a working balcony. Beneath the balcony there are steps leading to a door. Upstage is a view of the Champagne countryside with an old castle perched on a rocky summit. Between this upstage canvas and the stage itself is a small, vine-covered wall that encloses the garden. The wall has several gaps and a gate. At left, near the tree, is a table, two goblets, a pitcher, and two stools.

SCENE 1

ELIGIUS, WEATHERVANE

(As the curtain rises, Eligius and Weathervane are seated at the table next to the large tree, drinking.)

WEATHERVANE, *glass in hand.*

Besides that head cold you caught—or that caught you—did you have a good trip, my dear Eligius?

ELIGIUS, *gravely.*

Excellent, Monsieur Weathervane. (*He sneezes.*) Have a look! (*He takes a hatbox, a satchel, and an enormous encyclopedia off his lap and places them on the ground.*)

WEATHERVANE

Well then, to your health my dear son-in-law! (*They drink.*)

ELIGIUS

To your health, father-in-law. (*He sneezes.*)

WEATHERVANE

When I say "son-in-law," I know I'm getting ahead of myself a bit. But, my goodness, since the contract is to be signed tomorrow by King Dagobert himself, I don't see why I shouldn't allow myself this little indulgence.

ELIGIUS

Very true.

WEATHERVANE

I won't keep it from you: when you came here to the Champagne region for the first time three months ago to ask for my daughter's hand, you immediately won me over. (*As Weathervane speaks, Eligius takes a crown out of the hatbox and begins polishing it. Weathervane looks at it with surprise.*) What in the world is that?

ELIGIUS

This? It's Dagobert's crown.

WEATHERVANE

A replica?

ELIGIUS

Well, you know, the king was strapped for cash, so he pawned the real one to my aunt, and we replaced it with this one. I made it myself. It's plated.

WEATHERVANE, *with a smile.*

I see. So when the king puts on his crown, he's plated, too?

ELIGIUS, *gravely.*

In legal terms, we call that . . . (*he sneezes*) a play on words. (*They both stand and sit back down.*)

WEATHERVANE

Where was I? Oh yes. As a goldsmith, I used to be your boss. But now I'm extremely proud to see you established as an adviser to the king. By the way, when they speak, do French kings use the "royal *oui*"?[37] Anyway, Lucette was right to put up token resistance when I told her about your marriage proposal. And I did as you requested: I hid your titles and your qualities so she could love you for yourself. Believe it or not, she thinks you're a simple merchant, you, the great Eligius, prime minister of the kingdom!

ELIGIUS

In legal terms, we call that a . . . (*he sneezes*) a gimmick. (*They both stand and sit back down.*)

WEATHERVANE

She found you completely lacking grace and . . . youth. What's more, to coin a phrase that will become all the rage in the nineteenth century, she said that you completely lacked *moxie.*

ELIGIUS

Once we're married, I hope Mademoiselle Lucette will soon have a new . . . a new . . . (*he sneezes*).

WEATHERVANE

What? A newborn?!

ELIGIUS

A newly found love for me!

WEATHERVANE

Obviously. And when she came to talk to me about the songs Goldbrick put in her head, oh! That's when I became truly eloquent! My poor girl, I said to her, how can you possibly still have feelings for that scoundrel? A fraud of a poet! A clown with no promise who will forever be poor! He will never—and I mean never—be my son-in-law! Eligius is wealthy! Eligius pleases me! Lucette, my daughter, you will take the name of Madame Eligius! Speaking of Goldbrick, he is gone . . . on a trip to Maisons-Laffitte.[38] The coast is clear. We can rest easy! (*The sound of a horn can be heard in the distance.*) Ah! What a strange-sounding horn!

ELIGIUS, *walking to the gate.*

It's a valet announcing the arrival of the king. (*Exclaiming.*) The king! (*A loud noise is heard backstage.*) Oh no! The king has taken a spill!

SCENE 2

ELIGIUS, WEATHERVANE, DAGOBERT

(Dagobert, who rode a velocipede across the stage during Eligius's last lines above, returns, pushing his velocipede by hand. He places it upstage, center.)

Trio

ELIGIUS

Oh, he slipped off his wheels!

WEATHERVANE

And fell head over heels!

ELIGIUS and WEATHERVANE

Lord, we saw that you fell;
We both fear you're not well.

DAGOBERT, *checking himself.*

Well, it seems that my watch
Is the only thing botched.

And I just broke its glass
When I fell on my ass!
(*He springs to his feet.*)
Tra la la, tra la la,
Oh, I didn't get hurt and my body's intact.
Tra la la, tra la la,
Safe and sound, thanks to God! And look! No bones are cracked.

DAGOBERT, ELIGIUS, WEATHERVANE, *singing all together, drinking.*

Tra la la, tra la la,
Oh, he didn't get hurt and his body's intact.
Tra la la, tra la la,
Safe and sound, thanks to God! And look! No bones are cracked.

DAGOBERT, *to Eligius.*

You know what? I've had enough of your velocipede. It's too much. Since your damned invention nearly cost my life—bam!—you're no longer my prime minister. Give me your little notebook.

ELIGIUS, *takes a small registry from his satchel and hands it to Dagobert.*

Here you go, sire.

DAGOBERT

Hold it! I forbid using my noble titles. Call me Alfred. I'm here to leisurely enjoy a good feast at your wedding. I plan to stay completely incognito. Here, take my hair and put it with that crown. This wig could give me away.

ELIGIUS

Yes, sire.

DAGOBERT

Alfred!

ELIGIUS

Yes, sire.

DAGOBERT, *loudly.*

Alfred!

ELIGIUS

Right . . . Alfred.

DAGOBERT

That's it.

WEATHERVANE, *examining the velocipede.*

Eligius invented this machine?

DAGOBERT

My dear Weathervane, there is a long story behind this velocipede. Let me explain the circumstances surrounding its construction. It all starts about thirty years ago. My brother, Caribert, had a little amorous adventure in the village of Roche-Trompette with a certain boot seamstress named Bambochinette.[39] I don't know how well you know your history, but our father, Clotaire, was excessively rigid; by that I mean strict.

WEATHERVANE

Really!

DAGOBERT

Oh, it was extreme. One day he found out about his son's affair with the young Bambochinette, who, in the meantime, had become a mother to a young little lad named . . . what was it? Named . . . ? Eligius, what was the name of my brother's son?

ELIGIUS

Alfred, I have no idea.

DAGOBERT

Hand me my encyclopedia.[40] (*Eligius gives him the large encyclopedia.*)

WEATHERVANE

What's that book called?

DAGOBERT

An encyclopedia! It's extremely handy. You can find biographical information about historical figures from every country, information about mythology, modern and ancient geography, etc., etc. My whole life is in here from my birth to my death.[41] (*He opens the book and thumbs through it.*) If I don't know what to do—boom!—I look up "Dagobert" and right away I see what I did and then I do it! (*Scanning a page.*) Caribert . . . Caribert . . . Here it is! What? It doesn't include my brother's son's name. (*He closes the book and gives it back to Eligius.*) Oh well. No big deal. I'll keep going with my story. Are you keeping up, Weathervane?

WEATHERVANE

Yes, sire.

DAGOBERT

Alfred!

WEATHERVANE

Yes, sire.

DAGOBERT, *loudly.*

Alfred!

ELIGIUS, *to Weathervane.*

Call him Alfred. Come on!

WEATHERVANE

Right. Alfred.

DAGOBERT

That's it. So, what did Papa Clotaire do? The young woman lived about eight leagues away from the Armenonville Pavilion, where we were living at the time.[42] Since it was physically impossible to make the trek *pedibus cum jambis*—on foot—Papa Clotaire locked up the stables. Caribert had no way of getting a neigh-neigh for nightly visits to his love nest.

WEATHERVANE

Didn't he have enough money to buy himself a horse?

DAGOBERT

Are you kidding? Mama would give us only ten cents a week. So what do you think happened?

WEATHERVANE

That's what I want to know.

DAGOBERT

As an administrator, Eligius is clueless, but as a mechanic, he is even better than engineers from the Mulhouse Railway Company.[43] Eligius invented the velocipede to help my brother. Unfortunately, one night as he was going to see his beloved Bambochinette, Caribert forgot to brake during a descent, fell down, and broke his crown. I inherited the velocipede, of course. And that's the story.

WEATHERVANE, *inspecting the velocipede.*

It's very elegant.

DAGOBERT

Oh, it's very well made. The frame's tubes are hollow.

WEATHERVANE

But it doesn't look comfortable.

DAGOBERT

Hah! No.

ELIGIUS

That's true, it's not comfortable.

DAGOBERT

You agree?

ELIGIUS

Yes, I . . . (*He sneezes.*) I agree.

DAGOBERT

You've been a good boy, so I'll give back your little notebook. (*He gives it to Eligius.*) My friends, that little ride through the countryside has left me terribly famished. (*Eligius moves the velocipede offstage left.*)

WEATHERVANE

Alfred, I was about to check on dinner when you arrived. If you'd like to accompany me, I can show you our city's monuments on the way.

DAGOBERT

What is this city, by the way?

ELIGIUS

Troyes, in the Champagne region.

DAGOBERT

Champagne? Eligius, give me my encyclopedia. (*Eligius gives it to him. Dagobert searches.*) Champagne . . . Champagne . . . What are its culinary specialties? Ah, here they are: Champagne is known for slate, limestone, and the Marne River, etc., etc. But it is primarily famous for its white and red wines and for its sparkling wines known as champagne. (*Lucette appears on the balcony and listens in. Dagobert closes the encyclopedia and gives it back to Eligius.*) Yum. I hope we get to spend some quality time with a bottle of champagne.

WEATHERVANE

My dear Alfred, I was planning on bringing you an entire case!

DAGOBERT

A case? Well, in that *case*, come with us, Eligius. We'll have you carry it. You've always wanted to be in the *lap* of luxury. Now you can *lap* up whatever we don't drink![44]

ELIGIUS, *gravely.*

In legal terms . . .

DAGOBERT

It's what you call a play on words. And now, double time, forward, march! (*Lucette exits.*)

CHORUS, *sung by all the characters on stage.*

Tra la la, tra la la,
I didn't get hurt and my body's intact.
Tra la la, tra la la,
Safe and sound, thanks to God! And look! No bones are cracked.

(*They dance their way offstage.*)

SCENE 3

LUCETTE, followed by GOLDBRICK

LUCETTE, *alone.*

Papa just left with those gentlemen. Alas! (*She goes to the garden gate.*) I'm being forced to become Madame Eligius. Won't it be pleasant to spend my young life with a husband I detest? People may disagree, but I don't care. I say it's unfortunate I didn't get to choose my father, and it's unfortunate I won't be able to choose my husband. And Goldbrick isn't here. The ingrate must have forgotten about me. He probably has a new girlfriend by now. Ah! If I knew who, my little pink fingernails would teach her not to mess with my lover.

GOLDBRICK, *from behind the wall upstage.*

Psst!

LUCETTE, *surprised.*

Goldbrick!

GOLDBRICK, *jumping into view.*

Peek-a-boo! Here I am!

Duet

LUCETTE

Oh yes! It's my sweet love! Goldbrick, I'm overjoyed!
Come here, we're all alone.

GOLDBRICK

Lucette, I love you so!
Don't fear, I'm still your beau.
My love, now that I'm here, tell me you're not annoyed.

LUCETTE

I forgive you, my love.
You're a gift from above.

GOLDBRICK

I'm at your beck and call.

LUCETTE

But tell me, please tell me: Did you not get my note?
I wasn't sure at all.
Why did you come so late? Did you see what I wrote?

GOLDBRICK, *first couplet.*

When I got the news,
Lucette, I fell prey:
A case of the blues

Came quickly my way.
Despite my best work
I couldn't stay sage:
I soon went berserk
In a jealous rage.
I tried to think how
To turn things around
But found no good way
To rescue the day.
I wanted to die
Though my life's just begun.
My plan went awry,
I don't own a gun!
O goddess of youth
And goddess of love,
Please reveal your truth!
Send help from above.
Goddess, hear us pray.
We can't be undone.
Please show us the way
That we may be one.

GOLDBRICK and LUCETTE

O goddess of youth,
O goddess of love,
Please reveal your truth!
Send help from above.
Goddess, hear us pray.
We can't be undone.
Please show us the way
That we may be one.

GOLDBRICK, *second couplet.*

Then I thought I saw
Cytherea's young child.[45]

He filled me with awe,
He touched me and smiled.
The darkness of night
Was suddenly gone,
And to my delight
I saw a new dawn.
From Maisons-Laffitte
I left in a flash.
I came here, toot sweet!
I made a mad dash.
We can't stay apart,
I have to be near.
Lucette, my sweetheart,
Your true love is here!
O goddess of youth,
O goddess of love,
Please reveal your truth!
Send help from above.
Goddess, hear us pray.
We can't be undone
Please show us the way
That we may be one.

GOLDBRICK and LUCETTE

O goddess of youth
And goddess of love,
Please reveal your truth!
Send help from above.
Goddess, hear us pray.
We can't be undone.
Please show us the way
That we may be one.

(*Goldbrick places his beret on the table.*)

LUCETTE

Oh! I'm so happy and completely reassured. You understand that Papa has ordered me to marry someone else?

GOLDBRICK

Ha! I'd like to see him try to marry you!

LUCETTE, *worried.*

You'd like to see that?

GOLDBRICK

Umm, no. You see . . . it's just a manner of speaking. What will become of me? I'm a poor young orphan. If I'm separated from you, too, I don't know what I'll become. My rival must be wealthy to have persuaded your father. Either that or he's incredibly handsome. (*Eligius, carrying a case of champagne, appears at the garden gate and listens in.*)

LUCETTE

Not at all. He's ugly, he's old, he's awful, and I hate him as much as I love you.

GOLDBRICK, *embracing Lucette.*

My sweet little Lucette!

ELIGIUS, *outraged.*

Oh!

GOLDBRICK

But is he rich?

LUCETTE

Very. And like I told you in my letter, I don't see any way out of this sad situation. You'll just have to abduct me.

GOLDBRICK

That was my plan.

LUCETTE

I imagine that being ravished must be fun. On top of that, today's the perfect day for it.

GOLDBRICK

I'd go so far as to say that every woman with a bit of self-respect should be abducted at least once in her life.

LUCETTE

In that case, come back tonight at eight o'clock with a carriage, a dark lantern,[46] and a ladder. Clap your hands three times, and I'll come out onto my balcony. Then . . .

GOLDBRICK

But why don't we just leave right now? It seems—

LUCETTE

No! Papa might see us, and then all would be lost. Tonight he'll be asleep, and we'll have nothing to fear.

GOLDBRICK

Be ready at eight o'clock. I'll be right on time.

LUCETTE

Eight o'clock.

ELIGIUS, *aside.*

I'm going to warn the king. (*He exits.*)

GOLDBRICK, *embracing Lucette.*

Adieu, my sweet little Lucette.

LUCETTE

Goodbye, my little husband. (*As Goldbrick begins to leave, laughter, shouting, and rumblings can be heard offstage.*) Oh, my! What's that noise?

GOLDBRICK, *looking at the gate.*

There's a fight!

LUCETTE, *looking offstage.*

Wait. It's Monsieur Eligius.

GOLDBRICK

Eligius is a soldier?

LUCETTE

No, the other one. The old man.

GOLDBRICK

Wow! You were right. He's not attractive.

LUCETTE

And not courageous. He looks so sheepish.

GOLDBRICK

More people are joining in. They're trying to take a bottle of champagne from him. He's digging in his heels. They're pulling his hair. He's kicking them. (*Loud laughter.*) Oh! A bourgeois is coming to his defense.

LUCETTE

That's his friend. His name is Alfred.

GOLDBRICK

They're coming this way! (*Lucette and Goldbrick run downstage.*)

LUCETTE

Oh my God! If they catch us together, they'll tell Papa everything and we'll have no chance of success. Hide over there. (*She points behind the tree.*) I'll go back inside. (*She hurries into the pavilion.*)

SCENE 4

ELIGIUS, DAGOBERT, GOLDBRICK, in hiding.

DAGOBERT, *speaking with Eligius as they enter.*

I've always said you're a scatterbrain.

ELIGIUS, *his hand over his eye.*

Now more than ever. (*He puts the case down next to the tree.*)

DAGOBERT

What the hell happened? I was ahead of you and didn't see how it all started.

ELIGIUS, *takes his hand away revealing a bruised and swollen eye.*

As we left Monsieur Weathervane to order the meal, you . . . (*he sneezes*). You were walking ahead of me.

DAGOBERT

And you started running to catch up?

ELIGIUS

And I bumped the champagne case into a cavalryman who was chatting with a soldier, and kaboom!

DAGOBERT

He shoved you?

ELIGIUS

I held my ground. He got mad. I yelled. And then, whack! He put a fine point on the end of his sentence.

DAGOBERT

It's too bad he put that point right on your eye. In the name of Saint-Denis, my patron saint, you are unlucky.[47] But to return to the matter at hand, you said your rival was here, that you saw him with Weathervane's daughter, and that you know the time of her planned abduction?

GOLDBRICK, *hiding behind the tree, aside.*

Damn it! This monster was eavesdropping!

ELIGIUS

Here, tonight at eight o'clock, three claps.

DAGOBERT

We've got to figure out a way to head your rival off at the pass.

ELIGIUS

That's my thinking, too.

DAGOBERT

Your fiancée is pretty, you say?

ELIGIUS

An angel. (*He sneezes.*)

DAGOBERT, *aside.*

The encyclopedia says I've always loved women. I'll play a trick on my friend Eligius. (*He locks the pavilion door and puts the key in his pocket.*) Eligius, I have a plan.

ELIGIUS

So do I.

DAGOBERT
I have a plan, you have a plan. I'll wager my plan is better than yours.
ELIGIUS
I'll take that bet.
GOLDBRICK, *in hiding, aside.*
This is getting interesting.
DAGOBERT, *extending his hand to Eligius.*
Loser buys breakfast at the Porte Jaune.[48]
ELIGIUS, *they shake hands.*
It's a bet.
GOLDBRICK, *aside.*
What are they planning?
DAGOBERT, ELIGIUS

It is agreed,
We've sown the seed.
Now all we have to do is wait.
And soon we'll know
If this dodo
Can prove he's not a reprobate.

GOLDBRICK, *singing with Dagobert and Eligius.*

It is agreed,
We've sown the seed.
Now all we have to do is wait.
And soon they'll know
I'm pure as snow!
I'll prove I'm not a reprobate.

ELIGIUS, *looking around to make sure no one is listening.*
Can he hear us? I'll double-check he hasn't come.
(*He walks upstage to the wall.*)
DAGOBERT, *downstage, aside.*

I'll show him that I am not dumb.
My plan's the best, second to none.

Eligius may scheme and plot,
But in the end, he's far too taut.
While a man of my great stature
Need only follow his nature.
Since I know where Goldbrick will be
I'll dress like him and one, two, three.
Ha, ha, ha, ha!
And since I know the hour and sign,
I'll take Lucette and make her mine.
Yes, I'll beat Goldbrick to her perch
And leave the poor fool in the lurch.

DAGOBERT, ELIGIUS, GOLDBRICK, *Tyrolienne.*[49]

Tra la ee la la la ee la la
La la ee la la ee loo

(*Dagobert goes back upstage after taking the hat Goldbrick left on the table.*)

ELIGIUS, *downstage, aside.*

I'm going to prove that I'm not dumb.
My little plan will shock the scum
Named Goldbrick, who, without a doubt,
Is nothing but a worthless lout.
And when he climbs to meet Lucette,
I'll make sure that he is met
By soldiers from the regiment
And not his lover's sweet assent.
Ha, ha, ha, ha!
To catch this Romeo tonight
We've got to go, we must take flight.
We don't have time to take a break.
Tell the sheriff to stay awake!

DAGOBERT, ELIGIUS, GOLDBRICK, *Tyrolienne.*

Tra la ee la la la ee la la
La la ee la la ee loo

DAGOBERT, ELIGIUS

It is agreed,
We've sown the seed.
Now all we have to do is wait.
And soon we'll know
If this dodo
Can prove he's not a reprobate.

(*Dagobert and Eligius exit.*)

SCENE 5
GOLDBRICK, then LUCETTE

GOLDBRICK, *alone.*

If I understood what these two numbskulls are planning, I don't have a minute to lose, and I need to get ahead of them. (*He tries to open the door to the pavilion.*) Drat. It's locked. I'll have to get in through the balcony. But I need a ladder. (*He looks around.*) Whatever happens, it will be entertaining. I can already see Eligius locking up his friend, Alfred. Ha ha! I can't find a ladder anywhere. (*Calling out.*) Lucette! It's me, Goldbrick.

LUCETTE, *on the balcony.*

It's you! But it's not eight o'clock.

GOLDBRICK

It doesn't matter anymore. We've been found out. My rival knows our escape plans, so we have to leave early. Where can I find a ladder?

LUCETTE

Over there in the little greenhouse behind the garden.

GOLDBRICK

Perfect. Get ready. I'll be right back. (*He begins to exit.*)

LUCETTE

Oh my God!

GOLDBRICK

What is it?

LUCETTE

There! A man in a brown coat. Papa's coming home.

GOLDBRICK

No, it's Alfred coming to abduct you in my place!

LUCETTE

What are you talking about?

GOLDBRICK

Nothing but the truth.

LUCETTE

Here he comes. (*Dagobert enters upstage, wearing Goldbrick's beret.*)

GOLDBRICK

Go back inside. And this is very important: pretend you think he's me. Buy some time until Monsieur Eligius returns. Shh! (*Lucette goes back inside. Dagobert appears at the garden gate wearing an enormous coat. As he comes through the gate, Goldbrick deftly slides out behind him and exits.*)

SCENE 6

DAGOBERT, then LUCETTE

DAGOBERT, *alone.*

I really believe I was inspired. Such a bold idea! This is going to be such fun. My conduct is above reproach. The history of France teaches that I indulged in the extravagance that led to Solomon's ruin! One of my biographers even adds that my conquests are too numerous to list. If I'm giving in to another adventure, I'm only conforming to the idea historians have of me: I'm a lusty man! What else can I do? I'm playing my role, that's all! And no fear of crossing paths with Goldbrick; I'm half an hour ahead of him. Time to begin. (*He claps three times, then unlocks the door to the pavilion. Lucette appears at the door.*)

Duet

DAGOBERT

Come out, my dear, yoo-hoo!
It's Goldbrick here for you.
Come, my sweet, come this way.
Let me steal you away!

LUCETTE, *pretending to recognize Goldbrick.*

Is that your voice I hear?
Goldbrick, I'm glad you're near!

DAGOBERT, *embracing Lucette.*

Love and passion
Are a delight.
They light my light,
Are my fashion.
Sweet commotion,
Extreme pleasure,
Love's a treasure,
Alluring potion,
Alluring potion.

DAGOBERT, LUCETTE

Love and passion
Are a delight.
They light my light,
Are my fashion.
Sweet commotion,
Extreme pleasure,
Love's a treasure,
Alluring potion,
Alluring potion.

DAGOBERT

My dear, I love you so,
But now it's time to go.

LUCETTE, *worried.*

I don't want to embark
With that noise in the dark.

DAGOBERT

Those are other lovers singing their serenades,
Perhaps! But let's go!

SCENE 7

DAGOBERT, LUCETTE, GOLDBRICK, accompanied by several CLERICS who remain upstage

(Goldbrick is wearing a large coat; he has an enormous mustache, a helmet, and an immense sword.)

GOLDBRICK, *to the clerics who open their dark lanterns and illuminate the stage.*

Come over here, comrades!
I swear by all that is holy that I will make him pay.
That bumbling nitwit who, just beyond this chalet,
Nearly ran me over with his case of champagne!
Should he refuse to duel, we will knock out his brain.
Eligius, I'm told, is the name of that fool.
Our honor is at stake. We must kill that old ghoul!

GOLDBRICK, *whispering to Lucette.*

Mum's the word! I'm Goldbrick.

LUCETTE, *whispering to Goldbrick.*

Wow! What a disguise!

DAGOBERT, *aside.*

Damn!
It's the man who punched Eligius. I should scram!
He'd gut me like a fish in practically no time.
For Lucette's sake and mine, keep up this pantomime.

(During this whole scene, Dagobert makes it a point to hide his face from Lucette, who, for her part, makes Dagobert as uncomfortable as possible through her exaggerated curiosity.)

GOLDBRICK, *to Dagobert.*

Hello! Tell me, my good man,
Is Eligius at hand?
What a lout!

CHORUS

What a lout!

GOLDBRICK

Please tell me. I want to know.
Where did that nincompoop go?
Help me out!

CHORUS

Help me out!

GOLDBRICK

If you know his dwelling place,
Tell me; I need to slap his face.
He's a clown!

CHORUS

He's a clown!

GOLDBRICK

To keep my honor pure and clean
I have to skewer his little spleen
And take him down!

CHORUS

And take him down!

DAGOBERT, *brazenly.*

I don't know him, the little prick.
A simple man, my name's Goldbrick.

GOLDBRICK, *pretending to recognize him.*

What? You, Goldbrick? My dear old friend!
We met in Spain. How long's it been?
I think of you like my own kin:
All for one through thick and thin.
How are you, my long-lost twin?
How are you, my long-lost twin?

DAGOBERT

Fine, great, umm, yes, of course.

GOLDBRICK

It's really you at last!
We're going to have a blast.

CHORUS

Oh! This adventure's very taut.
The storyline's a Gordian knot.
We hope the puns aren't overwrought.
We'll try hard not to lose the plot!

DAGOBERT, *aside*.

Holy baloney! I've gotten myself into a real mess! And little Lucette doesn't seem to suspect that I'm not Goldbrick. But there's something strange about her. (*He sees the crate of champagne.*) Aha! The champagne! I'll get them drunk. (*Aloud.*) My friends, how about a drink of wine to celebrate the reunion of two old friends after fifteen years apart? (*Lucette goes into the pavilion and returns with champagne flutes.*)

GOLDBRICK

What a grand idea. We accept!

DAGOBERT

Then take a glass and pay close attention to the chorus so you can sing along. (*He distributes the glasses.*)

LUCETTE, *whispering to Goldbrick*.

What are you waiting for?

GOLDBRICK, *whispering to Lucette*.

For the soldiers to arrive. Once this gentleman is arrested in my place (*He points to Dagobert.*), your father won't be worried at all and we can leave in peace.

DAGOBERT, *aside*.

It seems nobody even notices I'm here. (*Aloud.*) Let's drink! (*He grabs a bottle.*)

ALL

Let's drink!

DAGOBERT, *first couplet.*

This bottle is so seductive:
Its silver neck, its curvy girth.[50]
Its contents can be instructive,
Inspiring love, and wit, and mirth.
You only have to drink one glass
Of this truly excellent juice
To let your inhibitions loose
And feel like you're riding first class.
It fizzes, it sparkles, it foams,
It's delicious come sun or rain.
When our imagination roams
We will always dream of champagne![51]

CHORUS

It fizzes, it sparkles, it foams,
It's delicious come sun or rain.
When our imagination roams
We will always dream of champagne!

DAGOBERT, *second couplet.*

King Dagobert is a good prince,
But sadly, when he gets plastered,
His folly makes everyone wince:
He puts his knickers on backward.[52]
This sparkling wine calms our jitters,
And its taste is to our liking.
So now to imitate our king,
Let's get drunk and put on knickers!

CHORUS

It fizzes, it sparkles, it foams,
It's delicious come sun or rain.
When our imagination roams,
We will always dream of champagne!
It fizzes, it sparkles, it foams,

It's delicious come sun or rain.
When our imagination roams,
We will always dream of champagne!

SCENE 8
DAGOBERT, LUCETTE, GOLDBRICK, CLERICS, WEATHERVANE, ELIGIUS, SOLDIERS

GOLDBRICK, *slapping Dagobert on the shoulder.*

My dear Goldbrick, I'm so glad to see you again. You're abducting the daughter of some mutton-headed bourgeois?

WEATHERVANE, *upstage.*

Muttonhead?

GOLDBRICK

Bravo, Goldbrick!

WEATHERVANE, *marching furiously toward Dagobert.*

Ah! You rascal! You can't deny that you were trying to run off with my daughter and drink my champagne. But Eligius warned me in time. Now I've got you, you reprobate, and you won't escape! Sheriff, do your duty!

SHERIFF

Which one is the perp?

ELIGIUS, *pointing at Dagobert.*

That's him.

DAGOBERT

Oh, please! I object!

ELIGIUS

Take him away! (*As the soldiers grab Dagobert, Goldbrick and the clerics make noise to drown out Dagobert's protests. The stage is in chaos. Everyone exits except Goldbrick and Lucette.*)

SCENE 10

GOLDBRICK, LUCETTE, WEATHERVANE, DAGOBERT, ELIGIUS, SHERIFF, CLERICS, SOLDIERS

(As Lucette and Goldbrick try to leave, all the other characters return, forming two rows at the garden gate, barring the exit. The soldiers have taken the clerics' lanterns and hold them open. Light fills the stage.)

ALL (except Goldbrick and Lucette), marching toward Goldbrick.

You shall not pass!
Don't even ask.
What's your hurry?
You're not worthy.
You shall not pass!

GOLDBRICK, *backed up against the front curtain.*

And why, good gentlemen, are you blocking my way?

ELIGIUS, *ironically.*

Friend, you certainly must know why you have to stay!

GOLDBRICK, *confidently.*

No, sir, why can't I leave? I am racking my brain.

DAGOBERT

Since you want and insist, allow me to explain.
We went out to the street.
After walking ten feet,
I pleaded with the soldiers,
I cried on their shoulders.
My true heartfelt appeal
Full of anger and zeal
Quickly changed the fervent
Heart of this old servant.
(He points to Eligius.)
Under a lantern's light
They see my face, my plight.

They then bow down to me,
Their hearts break with pity.
(*Everyone bows down to Dagobert.*)
Swiftly understanding,
At my firm commanding,
We rushed back here and looked,
And now your goose is cooked!

ALL (*except Goldbrick and Lucette*)

Swiftly understanding,
At my firm commanding,
We rushed back here and looked,
And now your goose is cooked!

GOLDBRICK, *removing his mustache, coat, and helmet, recites the following lines without music.*

Gentlemen, I have lost. So I must surrender.
I'm the lowly Goldbrick; I'm just a pretender.

LUCETTE

Oh my God! Is it you? Goldbrick, my one true love!

GOLDBRICK

Yes. And I accept my fate, my sweet little dove.
It's so unlucky to get found out when the victory was in sight.
Holy velocipede!

DAGOBERT

What did you say, young man?

GOLDBRICK

I said, "Holy velocipede!"

DAGOBERT

You know this type of vehicle?

GOLDBRICK

Of course. Papa had one.

DAGOBERT, *excitedly.*

And your papa, what was his name?

GOLDBRICK, *dejectedly.*

I don't know. I'm an illegitimate son.

DAGOBERT

What about your mother?

GOLDBRICK

Her name was Bambochinette.

DAGOBERT

A boot seamstress from the village of Roche-Trompette?

GOLDBRICK

Exactly.

DAGOBERT, *erupting.*

What?! Then you're my nephew!

GOLDBRICK, *reaching to embrace Dagobert.*

You're my uncle?

DAGOBERT, *stopping him.*

Wait. We have to be sure you're not playing me for a fool. Eligius, go get my velocipede. If you're really my brother's son, you'll be able to tell me what's engraved on the left pedal.

GOLDBRICK

Of course. The engraving reads, "My son has wheelie red hair." (*Eligius brings in the velocipede.*)

DAGOBERT

And what does that mean?

GOLDBRICK

It means, "My son has *really* red hair!"[53] See for yourself. (*Goldbrick shows them his bright red hair.*)

ELIGIUS, *gravely.*

Wheelie? In legal terms, we call that a pun. (*He gives the velocipede to a soldier.*)

DAGOBERT

Come here, nephew. (*They embrace.*)

WEATHERVANE

Gentlemen, I withdraw my complaint since Goldbrick is the king's nephew.

ALL, *surprised.*

The king?

DAGOBERT

Eligius, give me my hair and my crown. They're in the hatbox. (*Eligius gives them to him, and he puts them on.*) The final curtain is at hand. Since I no longer need to be incognito, I can wear my royal insignia.

ALL

Long live King Dagobert!

WEATHERVANE

I have completely changed my mind. Lucette, I allow you to marry Goldbrick.

ELIGIUS

But what . . . what . . . (*he sneezes*) what about me?

WEATHERVANE, *annoyed.*

You . . . you . . . you sneeze too much.

DAGOBERT, *to the soldiers.*

It would be unjust to make you come all the way here for nothing. Take a glass, and let's toast my nephew's engagement.

ALL

Let's drink!

It fizzes, it sparkles, it foams,
It's delicious come sun or rain.
When our imagination roams,
We will always dream of champagne!

(*Curtain.*)

CHAPTER THREE

NARRATING VELOCIPEDOMANIA

MANUAL OF THE VELOCIPEDE

Published in Paris in February 1869, the *Manual of the Velocipede* is the first full-length book to situate the *vélo* in the cultural spirit of the country that, thanks in large part to the Tour de France, has become synonymous with the bicycle. It includes instructions on how to ride and what size velocipede to choose, prices of components, advice on what to wear, a history of the velocipede, politics of the velocipede, predictions for the future, and even a short play. The longest chapters are dedicated to narratives about the velocipede: about the velocipede and love, the velocipede and war, the velocipede and fitness, the velocipede and gender roles. It is the first book to gather everything into a single volume, to tie all the loose threads of the French mania for the velocipede together, and to ground the velocipede in a national story, enshrining it in French culture in a way that provided two-wheelers real staying power. Replete with references to other literature, these narratives are grafted onto the great works of classical and French fiction, placing the modern, strange, mechanical velocipede in an established, respected, and human context. While unique for its treatment of the velocipede, the *Manual* is in good company in the nineteenth century, which saw a profusion of such generically diverse texts, including the craze of *physiologies* in the 1830s and 1840s and extending to the series of urban observations on which Walter Benjamin would base his Arcades Project: like them, the

Manual is an example of texts that Margaret Cohen has shown to be remarkable for their "heterogeneity."[1]

Take, for example, the chapter "Where a Velocipede Leads," which quotes a famous line from Pierre Corneille's 1637 play *Le Cid* and refers to the *blason*, a kind of poem popularized in the sixteenth century. The chapter also mentions the famous actor and librettist Jean Elleviou (1769–1842) as well as the possible discovery in 1863 of Leonardo da Vinci's remains. The chapter "Origin of the Velocipede" refers to the Trojan horse as a "Velocipede or wooden horse" before connecting the velocipede to the Bible, to works by Geoffrey Chaucer and Miguel de Cervantes, and to *Arabian Nights*.[2] Reaching into numerous literary genres and areas of scientific inquiry, the *Manual* sinks its hooks into the long history of the humanistic tradition, suggesting that this new machine and the mobility it affords are well grounded in the human experience—in how we love, act, and move. Unlike the other works about the velocipede from the period, the *Manual of the Velocipede* establishes a broad narrative of what it means to be a velocipedist. In short, the work translated here brands the velocipede in a way that would make it and its immediate descendant, the bicycle, a symbol of France, a sign of what it means to be French, and the embodiment of French cultural values.[3]

LE GRAND JACQUES AND THE VELOCIPEDE

The mixture of sport and literature in the *Manual of the Velocipede* also reflects the life of its author, Richard Lesclide, who wrote under the pseudonym Le Grand Jacques (and many variants).[4] Born in Bordeaux in 1825, Lesclide made his way to Paris, where he became an author, editor, and enthusiastic *véloceman*. He was friends with author Henry Murger; published editions by poets Paul Verlaine and Arthur Rimbaud; penned plays, an unfinished novel, and erotic stories; and, beginning in 1875, spent ten years working as Victor Hugo's secretary.[5] While working on these projects, Lesclide was an indefatigable promoter of the velocipede, launching a journal in 1869, *Le Vélocipède illustré*, that would reach a print run of seven thousand copies per biweekly edition before the Franco-Prussian War in 1870.[6] After the war and the Paris Commune in 1871, Lesclide kept the paper afloat for a short time under a new title, *La Vitesse* (Speed). After a nearly twenty-year hiatus, the journal reappeared in 1890, then continued under the editorship of Lesclide's wife, Juana (née Mélanie Jeanne Adrienne Ignard), from Lesclide's death in 1892 until 1899.[7]

In January 1869, Le Grand Jacques published *Almanach des Vélocipèdes illustré par un cheval sans ouvrage* (Velocipede Almanac Illustrated by a Jobless Horse) with the same editor (Librairie du Petit Journal) that would publish the *Manual of the Velocipede* later that winter.[8] In addition to a day-by-day calendar listing the hours of sunrise and sunset, the *Almanac* included a number of stories that Lesclide would revise for the *Manual*. As he wrote in the *Manual*'s preface, "The 1869 *Velocipede Almanac*, with a print run of several thousand, sold out in a few days. The public clamored for a new edition. Instead, we decided to gather everything that could possibly interest Velocipedists into a single practical and expanded volume. And yet we didn't abandon our flights of fancy; we kept some of the best articles from our *Almanac*."[9]

Lesclide also wrote a novel about the velocipede: *Le Tour du monde en vélocipède* (Around the World on a Velocipede). It first appeared in installments in his *Vélocipède illustré* before its publication in a single volume in 1869 and a final edition in 1872.[10] It tells the story of Jonathan Shopp, an aptly named wealthy American who plans to ride around the world with a "géante" called Victorine, a tall, robust woman he met outside of Paris who, Shopp determines, will have the strength for such a long trek. After gathering equipment and purchasing custom-made velocipedes, Shopp and Victorine set out from Paris on their adventure. The novel takes the form of the type of travel narrative that would reach its apogee several years later with Jules Verne's 1873 *Around the World in Eighty Days*.[11] *Le Tour du monde en vélocipède* ends as the two travelers reach the middle of Russia. Shopp rides on alone, unwilling to expose Victorine to the dangers ahead. And while the conclusion is ripe for a sequel, Lesclide never returned to the project. Following the table of contents at the end of the novel, the 1870 edition refers readers to Lesclide's paper, *Le Vélocipède illustré*, via a full-page ad, suggesting that the novel was, at least in part, written to drive subscriptions.[12]

EMILE BENASSIT

In order for the *Manual of the Velocipede* to appeal to the widest readership possible, Lesclide enlisted the help of illustrator Emile Benassit (1833–1902), who contributed to numerous newspapers and almanacs during the Second Empire.[13] Like the *Manual* itself, Benassit's work made inroads into all areas of French arts: he worked with the well-known caricaturist Etienne Carjat and illustrated for the novelist and editor Paul Duplessis. He also provided etchings for works by Alfred Delvau (more on this later), Provençal poet Paul Arène, novelist Alfred Sirven, and the prolific Charles Monselet, and completed a number of formal oil paintings, focused primarily on military scenes.[14]

Benassit provided all the illustrations for the *Manual of the Velocipede*. His challenge was to create a visual lexicon for the new machine and to follow the lead of Lesclide's *Manual* by making the velocipede both classic and modern: a natural fixture of contemporary French culture.[15] But the more pressing and practical challenge was to provide a detailed and accurate depiction of the velocipede when this vehicle could only be upright while in motion. Contemporary photographic processes required the subject to remain still for a lengthy exposure, meaning that early photographs of velocipedes are largely static studio pieces meant to capture riders of note, such as a well-known 1870 photo of Pierre Lallement cleverly seated on a propped-up velocipede. Benassit's solution was threefold. First, he underscores movement by depicting hats flying off, coats flapping in the breeze, and characters either on the brink of disaster or in dynamic positions. Figure 3.22 is a perfect example of this technique. In addition to coats flapping and hats flying, the bourgeois man leaning to his right and raising his umbrella to fend off the oncoming riders highlights the movement of the velocipedists. This particular sketch is set at night and the weak headlamps may have meant the riders did not see the pedestrians until it was too late (the shadow directly behind the couple confirms that the headlamps are the light source). We are left with the uncomfortable feeling that a crash is imminent, that the bourgeois on the left has done well to take action to protect his wife, and that she, for her part, has done well to protect her dog. The velocipedists' surprise heightens the sense of impending doom, the lead riders' eyes are comically exaggerated by goggles or glasses, and the riders both appear to have only one foot on a pedal, as if poised for a fall. The whole scene evokes movement, tension, and chaos. Second, rather than a smooth background with clear motion lines giving a sense of direction, Benassit gives the illusion of movement by sketching lines at jarring angles behind and around the riders. Figure 3.20 provides a case in point: the shadows and lines on the ground provide depth, but the lines in the background, grouped in various random angles, imply movement and tension as the female riders pass the more amply and conservatively dressed onlookers toward an unknown horizon away from the viewer—the riders, their mounts, and their outfits are carefully, almost statically drawn while the mishmash of angles in the background provide the illusion of movement and tension to mirror the tension required to keep a velocipede upright and progressing. Figure 3.23 provides yet another example of how Benassit conveyed the idea of the velocipede's movement, this time by depicting the velocipedist and his mount just after a crash. His arm and leg are raised, his hat teeters on the ground beside him, and the velocipede is safe and sound—and artfully drawn—between the fallen rider's legs and leaning on his chest. While the shadow lines on the ground provide depth, the curved lines rising in the upper left of the illustration hint at movement and perhaps even dust rising in the breeze as a result of our intrepid rider's fall. Finally, Benassit took care to draw spokes that radiate from the wheel's hub but do not extend all the way to the rim, thereby evoking the speed of blurred spokes, particularly near the outer edge of the wheel—figure 3.27 is a prime example of this technique.

In an attempt to normalize the new machine, Benassit took care to depict the velocipede as a natural addition to scenes of daily life. Figure 3.12, for example, is a fairly conventional depiction of a young couple kissing. Their velocipedes and the thatched and curved lines indicating movement neither disrupt their calm embrace nor trouble their romantic interaction. That their velocipedes are perfectly balanced and upright underscores the normalcy of the scene. In a similar fashion, figures 3.15 and 3.28 depict men out for a ride in the countryside. The soldier in figure 3.15 looks very much like a cavalryman and the velocipede could easily be replaced by a horse in a more conventional pastoral scene. In fact, Benassit frequently depicted horse-mounted military men in his oil paintings; his painting *Dragons à cheval*, for example, features a cavalryman in a nearly identical posture to the soldier in the velocipede illustration and the two works include the same menacing clouds in the background.[16] And were it not for the velocipede under the rider, figure 3.17 could be mistaken for a simple sketch of a pedestrian out for a stroll in the rain.

Benassit's most important endeavor may have been his depiction of women's fashion. Again, figure 3.20 provides an excellent example. Two women pedal under a tree and pass two other women standing to the side, one of whom is pointing at the velocipedists. The onlookers' long, heavy dresses contrast sharply with the light, form-fitting wardrobe of the cyclists. The riders also wear skullcaps topped with feathers that serve as ornamentation but also highlight the speed of the women as they propel forward on their velocipedes. The glance exchanged by one of the velocipedists with one of the pedestrians serves as an acknowledgment of what they share as women, but also, perhaps, of their differences: freedom and mobility on the left, layered immobility on the right. Figure 3.27 also deserves our attention. It depicts the character Pipette, a formidable cyclist who challenges gender norms of her time and who strikes a commanding figure in her story. Shown here as a modern Athena, Pipette wears a revealing outfit that highlights her strength, her femininity, and, most importantly, her speed—her skirt, hair, and headwear flutter behind her. Benassit's illustrations authorized daring feminine apparel and contributed to the growing sense that women's dress could and should evolve.[17]

THEMES

In what follows we analyze themes that appear throughout the *Manual*, both in the text and in Benassit's illustrations. Ranging from the aesthetic to the practical, the political to the erotic, they reveal how the emergent velocipede reflected a certain optimism present in French culture and society in the late 1860s. They also prefigure the cultural impact of the modern bicycle at the end of the century and into the 1900s.

The Velocipede and the Carnivalesque

Early representations of the velocipede conjured up the values of Carnival, the festive season that precedes Lent in the Christian calendar. As Mikhail Bakhtin explains, "All the symbols of the carnival idiom are filled with . . . the peculiar logic of the 'inside out,' of the 'turnabout,' of a continual shifting from travesties, humiliations, profanations, comic crownings and uncrownings."[18] Edward Muir sums up the carnivalesque neatly: "Carnival opens up the underworld of festive laughter and marketplace language. This underworld [emphasizes the] duality of the body, the distinction between, on the one hand, the material bodily lower stratum of ingestion and secretions and, on the other, the ascetic upper stratum of reason and piety."[19]

An early advertisement (figure 3.1), for both the Michaux velocipede company and a play that featured velocipedes, shows Carnival figures on velocipedes and directly ties cycling with the world of the Carnival.[20] The culture of the velocipede mirrored the ethos of Carnival by allowing women freedom of movement and authorizing them to challenge conventional fashions. It held out the promise of opportunities for men and women of different social classes to meet (though this

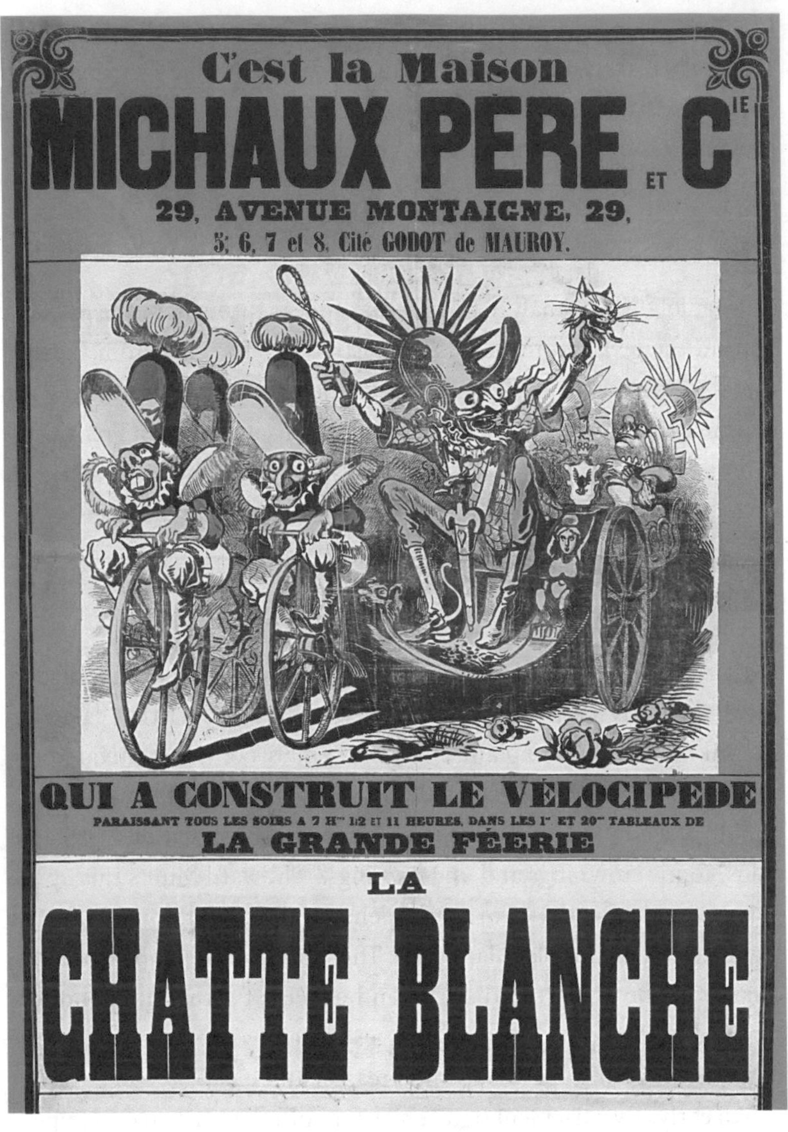

FIGURE 3.1. Paul Hadol, advertisement for the Michaux Père Company and the play *La Chatte blanche*, ca. 1869.

seems to have not been fully realized until much later), and it gave city dwellers the opportunity to leave their closed spaces and experience the air of the country—or at the very least of the Bois de Boulogne on the outskirts of Paris. It also allowed "crownings" of new champions, both men and women. In line with the upside-down nature of Carnival, the velocipede exercised the lower body, giving way to sweaty "secretions." And it lifted riders out of the mud and allowed them to fly over the ground under their own power, momentarily entering the ethereal realm of the sporting gods.

The *Manual of the Velocipede* underscores the carnivalesque spirit of the velocipede, featuring stories that elevate the erotic, describe travesties, and emphasize freedom from the conventional order. The stories are also filled with double entendres, the linguistic equivalent of the carnivalesque that brings high and low language together in a single word or phrase.

What's more, the *Manual of the Velocipede* was published by the press that also published *Le Petit Journal*, a popular newspaper known for covering Carnival—and especially the mid-Lent festival—extensively in its pages. Mid-Lent, also known as *la fête des blanchisseuses* (the festival of the laundresses), was a major event each year in nineteenth-century France. Characterized by cross-dressing, loud music, parades, and drinking, it elevated the lowest members of the feminine working class to the rank of queen for the day. In the *Manual* chapter titled "Velocipede Races," the lowly flower seller, Isabelle, watches women race on velocipedes on a track outside of Paris and immediately quits her job. She explains, "I'm done selling flowers. These Velocipedes have stirred my soul. I'm going to start training!" When a customer asks, "What has gotten into you?" she replies, "A vocation."[21] In other words, the velocipede, like Carnival, challenged traditional hierarchies and allowed women to aspire to something better, affording the promise—if only momentarily—of upward mobility even to women of modest means.

By describing and depicting the velocipede as enmeshed with Carnival, Lesclide and Michaux succeeded in weaving it into the fabric of a popular festival and grafting it into a centuries-old tradition associated with freedom, fun, and fancy. Though the velocipede was still expensive and out of reach for many readers of *Le Petit Journal*, it was less encumbered by the gender and class connotations of horseback riding and became, thanks largely to the *Manual of the Velocipede*, a symbol of liberation from class, gender, sartorial, and sexual strictures.

The Velocipede and Social Class

Lesclide's velocipede was something of a class paradox, but a paradox likely meant to attract the broadest possible readership to his manual and his newspaper. Before manufacturing processes became systematized later in the century, and before the bicycle would become a symbol of republicanism, the velocipede remained "popular among the upper class" (according to the *Manual*'s preface), since it was still relatively expensive.[22] The *Manual* recommends that serious new riders pay for an instructor while at the same time it presents itself as a how-to guide that can take the place of an instructor: "Our lessons summarize the science and advice of the most skilled professors."[23] The *Manual* recommends having a servant help during the post-ride routine and describes riding and racing at the exclusive Pré Catelan club in the Bois de Boulogne; but it also depicts women gaining a measure of freedom by riding the velocipede, shows a humble florist inspired to seek a new vocation by embracing the velocipede, and depicts a shopkeeper's daughter as a force on the velocipede. Though more aspirational than a reflection of true social change, the *Manual* presents the velocipede as both a pastime for the wealthy and a mechanism of upward mobility, a marker of exclusivity and a symbol of democracy.

Emerging Technology versus History, Man versus Machine

In a similarly paradoxical way, the *Manual* presents the velocipede as the epitome of technological progress on the one hand and as deeply rooted in history on the other. The preface explains, "After the coach, came the carriage; after the carriage, the railroad; after the railroad the Velocipede. Make no mistake about this most recent innovation; the wheels of progress continue to turn."[24] For Lesclide, the velocipede embodies the cutting edge of modern transportation, but he takes great care to connect the velocipede to the noble and ancient pastime of horseback riding. He refers to riders as *cavaliers*, or horse riders, and to the velocipede itself as a *cheval de race*, a thoroughbred. The illustration from the chapter "Where a Velocipede Leads" (figure 3.26) shows the velocipede held by a servant as though he were holding a horse, keeping it ready for its rider (most likely the woman in velocipede garb).

In a rather fantastical chapter titled "Origin of the Velocipede," Lesclide argues that "the horse . . . is certainly the founding idea, the seed, the type of a whole series of locomotive machines that have enriched humanity." He explains, "The first child to take a stick and run on it like riding a horse was the inventor of the primitive Velocipede."[25] This self-proclaimed "research-based" chapter goes on to draw a straight line from the horse to the velocipede using literary references: the Bible, Homer, *Arabian Nights*, and many others. It concludes with this claim: "You can see that the Velocipede is grounded in tradition and has its very own pedigree and nobility."[26]

Lesclide's presentation of the velocipede as both eminently modern and profoundly traditional is a well-tested approach for introducing something new by connecting it to the familiar, to the mythological, to the accepted. In his essay "Civilization Inoculated: Nostalgia and the Marketing of Emerging Technologies," Marc Olivier has shown how past innovators and promoters of new technology consistently used similar techniques (connecting the new to the familiar) to introduce and "socialize" their new machines—the microscope, the sewing machine, and the camera, for example.[27] This is precisely the strategy used by Lesclide in the *Manual of the Velocipede*.

In addition to clever marketing, Lesclide's bifurcated representation of the velocipede reflects a division in French culture of the 1860s, between traditional reactionaries and monarchists on the one hand and progressive positivists and republicans on the other. This split would become obvious in the early days of the Third Republic, when reactionary politicians attempted to restore a form of monarchy and seat the Count of Chambord (Henri V) as head of state. And the same rift would manifest itself again in the world of sport at the time of the 1900 Olympics in Paris, when the noble Baron Pierre de Coubertin's vision of restoring an athletic event from the distant past clashed with the republican organizers' desire to portray France as being at the forefront of technological progress.[28]

The *Manual* also mirrors the tensions between human and machine. It describes the body as a machine, comparing its multiple systems to the "broad keyboard of the human mechanism" and noting that exercise improves the "locomotive system."[29] But in the preface, Lesclide maintains that while the bicycle is indeed a machine, it is more elegant than the "brute and unintelligent" railroad. Instead, "The Velocipede represents personalized velocity emanating from man himself, reasoned rapidity bending to whims of the will, individual speed replacing collective speed, the affirmation that man is more powerful than steam."[30] In other words, the velocipede allows humans to master the rise of the machine, to submit the products of the industrial age to the will of the individual.[31] The velocipede, a machine, paradoxically liberates humans from the machine age; it allows the human machine to tame the industrial machine.

The Velocipedic Erotic

The *Manual of the Velocipede* is filled with sexual allusions and double entendres. In the chapter titled "Marriage on a Velocipede," purchasing a new velocipede is conflated with falling in love. As the narrator begins to fall in love with another velocipedist named Dorothy, he claims, "Happiness was in my legs."[32] He later describes the female object of his affection in these terms: "The young enthusiast's blouse rose and fell, quivering like a trapped dove, and I dreamed of spheres and snowy mountain peaks."[33] In the same story, the narrator speaks of purchasing lanterns from the shopkeeper whose daughter he loves and then adds, "That is, if the term 'lantern' is not risqué."[34] This extra clause calls the reader's attention to the double entendre: the word "lantern" is nineteenth-century French slang for vagina. In a chapter outlining the velocipede's history, the reader learns that the velocipede "is derived from the rod," which "conducts, directs, and governs."[35] In the chapter "A Velocipede's Lover," the velocipede functions as a synecdoche of its male rider, who "managed to avoid any piqued elation in the area of his member, where he experienced sweet sensations in the presence of Clémence."[36] Though this is perhaps more a single entendre, it overtly calls attention to the sexually charged language that runs throughout the work.

This use of language situates the *Manual* alongside Parisian theatrical productions from the same period. *Comédies vaudeville*, operettas, opéras bouffes (like *Dagobert and His Velocipede*), and boulevard plays (like those written by Eugène Labiche, Georges Feydeau, and many others) remained incredibly popular in the 1860s and 1870s. These productions almost always centered on love triangles and featured numerous quid pro quos and double entendres. Lesclide himself wrote several plays in 1876, including *Une maison de fous* and two one-act plays about the stock character Pierrot: *Pierrot en prison* and *Le Premier Duel de Pierrot*. The short theatrical play he penned for the *Manual of the Velocipede*, "Velocipede Races," prefigures his other plays and features its fair share of flirtatious titillation.

Such pulsating eroticism from Richard Lesclide is not surprising, given that he wrote several erotic texts. His *Contes extravagants* (1886) is a collection of short erotic stories, though Besnier's illustrations in that volume are more suggestive than Lesclide's fairly circumspect narratives. Lesclide's novel *La Femme impossible* (1883) centers on a woman who, for a reason never fully explained, cannot make love and, despite her desire, dies a virgin. But alongside the *Manual of the Velocipede*, Lesclide's novel *La Diligence de Lyon* (The Lyon Stagecoach, 1882) remains his most enduring work. The most fully articulated version of a legend that had circulated in libertine literature—Paul

Verlaine mentioned it in a poem in the collective volume *Album zutique* (1871)[37]—it relates the story of a nobleman, Lord Algerton, who late one night is offered the "Lyon Stagecoach" by a woman in the street. He then spends the rest of the novel trying to figure out the phrase's meaning. It becomes clear that it is a sexual act or position, but it is never clearly explained in detail. At the end of the novel, after much travel, many inquiries, and multiple false hopes, Lord Algerton finally meets the woman, a marquise, who can show him the Lyon Stagecoach. The narrator, preoccupied with another woman, leaves Algerton but is soon alerted that something sinister has happened to his friend. The narrator and his companion are led to "a padded room, furnished with strange instruments," where they see Algerton and the marquise "lying on cushions in disarray, pale and immobile as if struck by lightning."[38] The Lyon Stagecoach, it appears, led to the lovers' immediate deaths.

If Lesclide coyly avoided describing the act, his contemporary, Alfred Delvau, was less restrained. In the *Dictionnaire érotique moderne* (Modern Erotic Dictionary, 1864), Delvau provides the following definition:

> The Lyon Stagecoach. It is one of the rarest and most curious positions. Numerous followers of Venus have died without ever knowing it. This is because, in order to execute it, one must find a woman that possesses two rare qualities: [enthusiasm and freedom from prudishness]. When a woman is both passionate and liberated, she finds a man who pleases her in every way; she strips him stark naked, lies him on a bed with cushions under his head and lower back, and naked herself, she straddles him like a horse, skewers herself on the natural pivot point, in other words his member. Then she acts like the coachman on one of the horses of the old Lyon stagecoaches. Leaning forward on her lover's shoulders she moves forward, riding him. . . . She moves more and more vigorously as if the stagecoach were covering bumpy ground. . . . The woman then falls as though dead in her lover's arms.[39]

Upon reading this definition by Delvau, we can theorize a connection between Lesclide's novel *La Diligence de Lyon* and his *Manual of the Velocipede.* In both texts, whether in reference to a velocipede or a lover, "riding" leads to sexual encounters. In both cases, women are liberated and take initiative, independently controlling what they are riding. Whether going over bumpy ground on a boneshaker or riding her lover, the woman's gratification is paramount and takes priority over the man's. What's more, horseback riding—especially when the rider straddles the horse—is traditionally gendered as male, but for Lesclide, horseback riding, sex, and the velocipede, in carnivalesque fashion, flip the gendered hierarchies, putting the woman in charge of her own path and her own pleasure.

It is worth pointing out that Delvau published the majority of his erotic works beyond the reach of French censors, either in Switzerland or in Belgium; there, with fewer restrictions on tone and content, he could be more direct and explicit in his writing.[40] He did not exclusively write salacious works that had to be published outside France, however; the author of the *Dictionnaire érotique moderne* collaborated with Emile Benassit, the illustrator of the *Manual of the Velocipede,* to publish a book in 1866 (the same year Gustave Courbet completed his painting *L'Origine du monde*) that detailed bohemian life in Paris hour by hour: *Les Heures parisiennes.* Benassit's etching for the chapter titled "Midnight" (figure 3.2) leaves very little to the imagination and can be read as willfully pornographic.[41]

Three years later, Benassit produced a similar sketch for the *Manual of the Velocipede* (figure 3.3). While this second image is less overt and the woman's dress is absent, it nevertheless retains the erotic implications of Benassit's earlier etching. In a form of artistic synecdoche, the woman's clothes, including the pants she would have worn while riding the velocipede, appear to be stacked with the man's clothes on the chair, and the heater has been shifted to the right side to balance out the composition. The canopy opening looks a bit less explicit, but the presence of the large leaf above it calls to mind the many paintings that feature Adam and Eve (and other classical figures) covered

FIGURE 3.2. Emile Benassit, "Midnight," in Alfred Delvau, *Les Heures parisiennes*, 1866.

FIGURE 3.3. Emile Benassit, "Sometimes, everything just *falls* into place," *Manual of the Velocipede*, 1869.

with leaves to protect their modesty. Benassit's illustration works with the chapter that precedes it ("Where a Velocipede Leads") to underscore the velocipedic erotic and to situate it and his other sketches in a visual network that hearkens back to Delvau and connects the velocipede to classical erotic art. The erotic may be more disguised in the *Manual of the Velocipede* than in Lesclide's or Benassit's other works, but it remains nevertheless a significant theme hinted at throughout.

The Velocipede and Gender Roles

Though the *Manual of the Velocipede* often traffics in clichéd sexism and reinforces gendered stereotypes, it nevertheless promotes a modest form of early feminism, at times granting women an equal role to men and suggesting that the velocipede opens opportunities for women to pursue their own course. In this respect, the rise of the velocipede's popularity mirrors the rise of feminist movements in 1869 France. In the late 1860s, French women began to organize and push for meaningful changes to the Civil Code. In 1869, the feminist journal *Le Droit des femmes* published a manifesto signed by a number of women, including the authors Maria Deraismes and André Léo (née Victoire Léodile Béra), two women who had spent years writing in favor of women's rights, women's education, and female labor groups. The journal's "objective [was] to mobilize public opinion in favor of the 'civil rights of women,' access to a secondary and university education, the right to work, and equality of salaries."[42]

The *Manual of the Velocipede* is by no means a full-throated defense of feminism. Rather, like French society as a whole in 1869, it paints an inconsistent picture of women's roles. The chapter "Velocipedes of the Pré Catelan" mentions races for women, then immediately discusses the female velocipedist's "attractiveness."[43] In a discussion of the benefits of exercise, the text explains without any irony that a "woman is a big and admirable child" who must maintain her strength and flexibility in order "to conceive and give birth." The short play included in the *Manual*, "Velocipede Races," focuses on a women's race for which men are props whose only function seems to be to cheer the women on. But even here the character Bernard can do little more than discuss the velocipedists' legs. "A Velocipede's Lovers" is a story of love and riding in which the male character is a personified bike (referred to as "the Velocipede") while the woman, named Clémence, is a more developed character than her masculine counterpart. In a chapter devoted to velocipede fashion, men's clothing is described briefly and without judgment; however, the section on women's clothing is much longer and contains prescriptions for what women should and shouldn't wear.

Lesclide captured a popular sentiment that associated the velocipede with eroticism and with a limited measure of gender equality. Two 1869 images from *Paris-Caprice* (printed just before the *Manual of the Velocipede* was published) reflect liberation from conventional gender roles spurred by the velocipede itself. Figure 3.4, titled "Line Drawing of Paris" and accompanied by the caption "Proud whinnies, oh my velocipede!" shows velocipedists speeding away from horses, implying the velocipede's superiority. In addition, the drawing evokes the carnivalesque in all its splendor and chaos, with people of all different ages and social classes (suggested by the multiple styles of dresses and hats) stumbling, cheering, and riding. Finally, we can see both men and women riding together, speeding away from the soldiers on horseback. Significantly, a woman, second in the line of velocipedists, waves to the other riders, leading them forward as if into battle. The velocipedists are poised and resolved, while the horsemen are struggling to control their animals, creating disorder around them, and the horse right in the middle of the sketch is prominently displaying its hindquarters as it lurches wildly toward spectators. In this image the velocipede has effectively replaced the horse as the more civilized, more respectable, and most democratic mode of transportation. Where only men are shown on horseback, women ride alongside men on velocipedes, with one even leading the group forward as a crowd cheers them on in the background.

"Season of Abductions" (figure 3.5) underscores the eroticism and also the gender equality associated with riding the velocipede as men

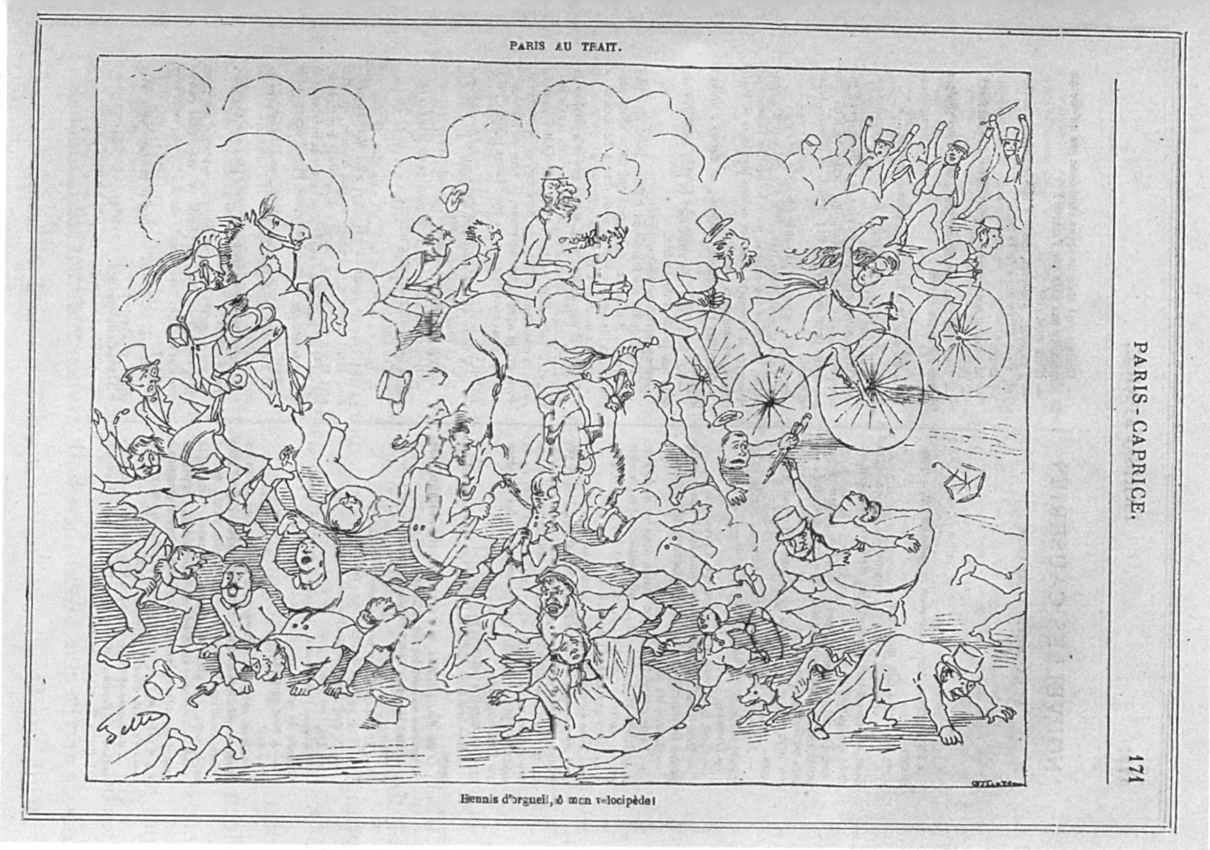

PARIS-CAPRICE. 171

PARIS AU TRAIT.

Hennis d'orgueil, ô mon vélocipède!

FIGURE 3.4. “Proud whinnies, oh my velocipede!” *Paris-Caprice*, February 20, 1869.

FIGURE 3.5. Albert Robida, "Season of Abductions," *Paris-Caprice*, March 6, 1869.

hold women and women hold men across their laps. The caption reads, "This type of exercise is becoming more and more popular: I abduct, you abduct, he or she abducts, etc.—You can clearly see that spring has sprung!" This caption implies rote memorization, a conventional method for learning French verb conjugations; the new machine, like its new lexicon, is available to all pronouns and to anyone willing to learn. Much less boring than memorizing verb endings, though, the velocipede offers the kind of mobility, freedom, and pleasure that will knock your hat off. Ride or be ridden: while it echoes the many paintings of the abduction of the Sabine women (most notably in France by Nicolas Poussin and Jacques David), here both women and men are "abducted."

While the *Manual of the Velocipede* is imbued with a number of gender stereotypes, it is also a progressive text, predicting that the velocipede will serve as an instrument contributing to greater agency for women, giving them increased individual speed and autonomy and enabling them to map their route, literally and figuratively. And while brief moments of a single woman's liberation were sometimes overtaken by the normalized gender roles of conventional marriage (see "Velocipede Races," for example), the *Manual*'s prediction would increasingly bear fruit, in fits and starts, over the following decades. Describing the influence of cycling on women in 1890s Victorian England, Patricia Marks concludes, "No other individual sport seemed to further the women's movement so radically. . . . This revolutionary traveling machine changed patterns of courtship, marriage, and work, to say nothing of transportation; it altered dress styles and language, exercise and education."[44]

CONCLUSION

As a whole, the *Manual of the Velocipede* presents readers with paradox after paradox. The velocipede is at once the great symbol of modernity and deeply rooted in human history; emblematic of society's elite, it nevertheless offers hope of upward mobility for the working class; it is instrumentalized to simultaneously liberate and objectify women; and, finally, it is an industrially produced machine that allows riders to escape the industrial world. Through these contradictions, it reflects the paradoxes of social and political life of late 1860s France. While presenting and reveling in these paradoxes, the *Manual of the Velocipede* remains the foundational text that establishes the figurative value of the velocipede and its descendant, the bicycle. The associative symbolic tropes set forth in the *Manual* have evolved over time but have nevertheless remained part and parcel of the bicycle's mythology to this day.

We have placed the illustrations by Emile Benassit (photographed by Brigham Young University's Digital Imaging Lab) approximately where they were included in the original. Figure 3.20 (by Dufaux) does not appear in the original text but is included for illustrative purposes. In the original, the illustrations are not captioned; we have added captions that help link the image to the accompanying text.

MANUAL OF THE VELOCIPEDE

By Le Grand Jacques
Illustrations by Emile Benassit
1869

MANUEL

DU

VÉLOCIPÈDE

PUBLIÉ PAR LE

GRAND JACQUES

ILLUSTRÉ PAR

ÉMILE BENASSIT

PARIS
LIBRAIRIE DU PETIT JOURNAL
—
1869

FIGURE 3.6. Cover of *Manuel du vélocipède* (*Manual of the Velocipede*), 1869.

PREFACE

FIGURE 3.7. Velocipedist.

The Velocipede is a sign of the times.[45]

After the coach, came the carriage; after the carriage, the railroad, after the railroad, the Velocipede.

Make no mistake about this most recent innovation; the wheels of progress continue to turn.

The railroad is undoubtedly faster; but it is mechanical, brute, and unintelligent.[46]

The Velocipede represents personalized velocity emanating from man himself, reasoned rapidity bending to whims of the will, individual speed replacing collective speed, the affirmation that man is more powerful than steam.

Style makes us human, said a great naturalist.[47] What could be more intimate than the relationship between an individual and the Velocipede he rides?

Just as Malibran channeled her soul through her voice, man channels his legs through the wheels that propel him forward.[48]

Thus, the popularity of the Velocipede is more than a trend or a sport; it is a fever.

This horse of wood and steel fills a void in modern existence; it fulfills not only our needs but also our aspirations.

The 1869 *Velocipede Almanac*, with a print run of several thousand, sold out in several days.[49]

The public clamored for a new edition. Instead, we decided to gather everything that could possibly interest Velocipedists into a single practical and expanded volume.

And yet we didn't abandon our flights of fancy; we kept some of the best articles from our *Almanac*.

In the words of the poet, "Utile dulci."[50] It is not enough to instruct people; one must also entertain them.

This is why, alongside serious articles, we have inserted humorous stories, illustrated by a talented artist.

This little book is not only a typographic gem but is also a well-argued manual of the Velocipede, the most reliable and most precise guide for disciples of this new form of locomotion.

In just a few pages, our lessons summarize the science and advice of the most skilled professors.

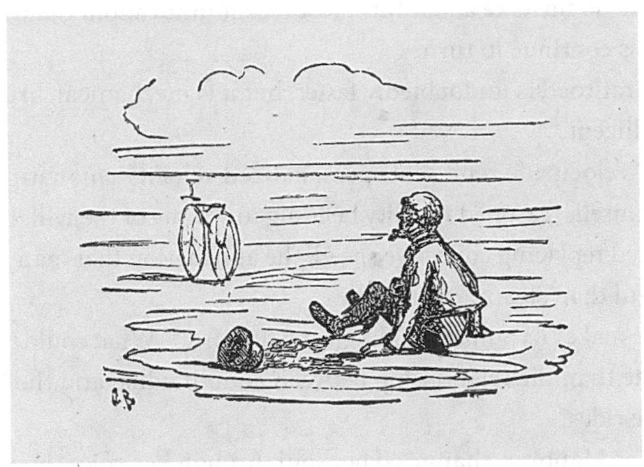

FIGURE 3.8. Contemplating a velocipede.

Questions about the Velocipede are thoroughly answered in detail from theoretical and practical perspectives.

In our opinion, the Velocipede is more than a passing fad. Even though it has quickly taken hold in modern life and has generated a growing sense of enthusiasm, that does not change the fact that it is immensely practical. This is proven by the fact that, even as it becomes more popular among the upper class, the government and its agencies use it for special projects.

The Velocipede will certainly remain. Perhaps one day we will study the influence of this new vehicle on the future, and our nephews will declare, parodying the famous phrase of Buffon, *The Velocipede is man's noblest conquest!*[51]

L. G. J.

THE ART OF THE VELOCIPEDE

FIGURE 3.9. Velocipede.

When a new word enters into the French language, it brings with it a certain number of related declensions and other new words that can be derived from it.[52]

At present, it is generally agreed to say "Velocipede" and "Velocipedist." The verb designating the action of movement seems harder to figure out. Should we say "Velocipate"? That word is far from the most graceful, but finding a better one is easier said than done.

Even this chapter's title isn't satisfactory. After all, we say, "The Art of Equitation" and not "The Art of the Horse." We had thought of "The Art of Mounting a Velocipede," but getting up on one isn't enough: the rider has to make it move. In the end, we thought it best to reduce our title to its simplest form.

While we wait for the dictionary to be enriched with specialty words to fill in the gaps, let us step away from such preoccupations, leaving behind words in favor of ideas.

~

The Velocipede is a means of locomotion that is in fact is not entirely modern. One has only to consult Bachaumont[53] to find, in the eighteenth century, hobbyists traveling on chariots, moved by springs that were pressed by the hands or the feet.

Going back further, there are rudimentary versions of Velocipedes in scientific essays from the fifteenth and sixteenth centuries, as a number of documents available in the Royal Library can attest.

The Velocipede was not entirely unknown in Antiquity.

In the frescoes of Pompei, for example, there are winged genies, straddling a two-wheeled baton: What simpler definition of today's Velocipede could there be?

Images that are analogous, although more rudimentary, are evident in Egyptians' hieroglyphs.

Finally, we could trace the Velocipede all the way back to the earliest moments of mythology, since Fortuna, on her Wheel, is obviously riding a Velocipede in all its perfection.

~

In reality, bluntly, the Velocipede is a simple improvement of the wheelchair.

We hope our readers actually consider what we are putting forward, instead of being surprised by it:

In its basic movement, doesn't the Velocipede purely and simply replace the force of walking with a force applied to the motion of a wheel?

The two machines are therefore of the same family.

—What does one gain by substituting the Velocipede for walking?

This is easy to calculate.

Suppose that we are on flat terrain paved with asphalt or macadam.

Walking, a man moves his own body, which science tells us has an average weight of 65 kilograms, with a speed of 1m50 per second—a force expenditure of about 100 kilogram-meters.*

(I apologize for using this barbarous term, which is impossible to avoid.)

On the other hand, studies of the ratio of "pulling effort to driven weight" have established that traction on a good, flat road can be evaluated to one-hundredth of the weight of the load.

However, we do not necessarily win ninety-nine times out of a hundred by using the Velocipede:

—First of all, because the weight of the Velocipede is added to that of the rider, and so it reduces the benefit by the same amount.[54] We thus arrive at a necessary effort of two kilogram-meters.

—Next, because this effort increases in direct proportion to speed. Now, if we fix walking speed at 1m50 per second, and the speed of a Velocipede reaching six meters per second, the ratio of one to four will bring the amount of force needed to eight kilogram-meters.

—Finally, because the Velocipede is at a disadvantage during climbs and along rough ground. The ratio of one to one hundred between drag and weight changes rapidly, and if the slope becomes steep or the ground too uneven, the expenditure of force applied to the machine is able to exceed that of a single step.

On the other hand, it is true that on descents the Velocipede often goes alone, and sometimes even too quickly, so much so that the rider has to moderate the speed by using the brakes.

But, limiting ourselves to the hypothesis that we have adopted of the uniform, flat terrain, we can confidently set at nine-tenths the

* The term *kilogram-meter* is used to denote the unit of force necessary to raise, by one meter in one second, the weight of one kilogram.

reduction of the traveler's effort when he replaces his legs with a two-wheeled Velocipede.

⁓

The Velocipedes that surround us are more or less perfected.

However, they come from a very simple machine, called the Celerifere, known for sixty years and which the Americans use quite widely.[55]

It is made of two free wheels, with no transmission of movement between them, connected by a bench low enough that the rider's feet can touch the ground, thereby providing support and ensuring balance.

⁓

"The Celerifere," explains Bescherelle, "is a kind of wooden horse on two wheels on which one balances while providing movement by pushing off with the feet."[56]

The Celerifere dates from the beginning of this century; it was used, under the First Empire, by some members of the administration, and it can be seen in the caricatures of the time. Forgotten during the Restoration, this basic machine seemed to return to favor in 1830. Some rural letter carriers then made use of it, but they ended up abandoning it because it was not able to withstand harsh winters.

The Celerifere, moreover, had rather serious inconveniences, not to mention that it required a rather high expenditure of force. The encounter of the foot on the ground was often rough and produced annoying jolts; in addition, this regular friction wore through shoes quickly. It is therefore unsurprising that the Celerifere was abandoned; but it is regrettable that at the time there was not a sufficiently ingenious mind to add the improvements that have recently made it so successful. The Celerifere was the egg; the Velocipede is the bird. The incubation period was long.

To the inquiring minds who would like to go back to the previous century and look for "The Origin of the Velocipede" in the mists of time, we recommend a humorous article that we publish with this very title, later in this volume.

Today's two-wheeled Velocipedes reach twice the speed of Celeriferes, with a more consistent pace.

But since the only contact with the ground comes in the form of two narrow wheels in a straight line, it means that the rider must stay upright as if on a tightrope, and he runs the risk of losing his balance.

Although we shouldn't exaggerate.

The fulcrum is not as cramped as it might sound.

The front wheel, controlled by a lever, can be moved to the right or left, so as to enlarge the polygon, in the middle of which is the center of gravity. In addition, the speed while in motion helps to keep the rider directly above the path of the two wheels.

But this creates a vicious circle:

To keep your balance easily, you have to go fast. But to go fast, you have to be in balance first. How to square that circle?

Experienced Velocipedists start by giving the machine a vigorous push while holding onto its handlebar.[57] They follow this momentum and climb up without stopping, adding to its speed. Their feet quickly find the pedals and are moved by them in a motion that at first they follow and then create as they keep it going. The steed is off, from then on obeying both feet and hands.

Less-practiced Velocipedists begin by familiarizing themselves with the older and simpler method of the Celeriferes, as has been discussed. They start by pushing the ground with their feet, in a series of steps; once they have attained the speed necessary to ensure their balance, their feet move from the ground to the pedals.

The question is whether one has legs long enough to partake in this exercise. For that, it is sometimes necessary to give beginning Velocipedists a little extra room. The diameter of the big wheel varies from eighty centimeters to roughly one meter ten centimeters; it is important to start with smaller diameters and work one's way up, keeping the rider's size in mind.

∽

One isn't born a full-fledged Velocipedist, but it is possible to have a particular aptitude for it or, alternatively, to be exceptionally clumsy. The students in this last category should not despair for the future; they just need to take a few days to put aside their pride, to steel their resolve, and to put themselves in the hands of a good teacher, who will quickly transform them.

This teacher is always useful, and we certainly do not recommend that people try riding without one. With some good guidance, an excellent gymnast will become a Velocipedist in just a few hours; that is the exception more than the rule.

A teacher always gives useful advice and helps beginners avoid the hesitations and the first falls that can be dangerous.

By taking courses at the Michaux Manège or the Paz Gymnasium, one can become a Velocipedist in every sense of the word, and after five or six lessons, one can boldly venture out onto the boulevards or the Champs-Elysées.

∽

The two-wheeled Velocipede is the only one that is truly worthy of the name. It is a purebred horse, while the three-wheeled Velocipede is a small coach.

It is true that with the three-wheeled variety one no longer need worry about balance—the rider sits upright on a kind of armchair—but gaining traction is more difficult and riding requires twice as much effort. Instead of contributing its own natural motion to the machine's movement, the body sits heavily on the seat.

It is more common for these kinds of cars to have twin engines.

Allow us to repeat that earlier point: between the two- and three-wheeled Velocipede is the kind of difference that separates a racehorse from a carriage horse.

At the end of this volume, readers will see that an important company, which is a leader in the industry, has very simply solved the problem by allowing riders to put two or three wheels, as they like, on the same Velocipede: they created one Velocipede to fill two distinct purposes.

Here is a summary of tips for the two-wheeled Velocipede:

Start off with a very low Velocipede, so that your feet touch the ground and keep you in balance.

Go faster and faster, with the aid of simple pushes of your feet.

As much as possible, for these first attempts, you should be on gently sloping ground, or at least horizontal ground, either paved or on parquet.

Place your hands on the ends of the handlebar, your body leaning back, your arms outstretched, your eyes looking straight ahead—for the Velocipede does not tolerate distractions, and only experienced riders can afford to greet friends whose paths they cross.

As you progress, and once you get up to a good speed, lift your feet off the ground and use your handlebar to help keep your balance.

By doing so, you will grow more comfortable going slowly, and you will gain confidence in the machine's stability. With your legs on each side of the frame, you will be ready to prevent a fall or at least to reduce its severity should it happen.

When you feel ready, press down on the pedal with your right foot while skimming the ground with your left foot to get used to the pedal's movement.

Repeat the exercise with the other foot.

When you are familiar with this way of riding, place both feet on the pedals one after the other and throw yourself into the exercise with confidence.

We have not described the mechanics of the handlebar, because people understand its function instinctively and typically use it well. It acts as a pendulum that sets the center of gravity according to the needs of the road. It draws the front wheel to the side of the hand that pulls it, and therefore tilts the whole system to the same side. If the rider feels he is leaning to the right, for example, an instinctive movement causes him to grab and hold firm with the right hand; the result is a movement of the handlebar and a slight correction on the right, which sends the machine toward the opposite side via centrifugal force.

Beginners and clumsy riders are easily identified by the handlebar wavering perpetually in their hands. At every moment they fear losing their balance, they constantly undermine their own forward motion, they weave a sinuous path, and they sometimes end up losing the balance they had been so eager to keep.

For an agile and somewhat experienced student, no serious accident is to be feared. The feet, which are in no way attached to the pedals, can leave them in the blink of an eye to regain contact with the ground and prevent a fall.

As we have said, the acceleration of movement restores and maintains balance. The center of gravity of a rolling system, racing at high speed, is less likely to deviate from the line drawn by the wheels: a line approximately two centimeters wide, on each side of the center of gravity. The momentary loss of equilibrium is at every moment redeemed and compensated, and the rider, hurtling at full speed, can cross large expanses, in a nearly straight line, without the aid of the handlebar.

When the Velocipede's momentum is decreased, balance becomes more and more unstable, for the reason that is the inverse of what we have just explained.

Only the handlebar can keep the slowly moving rider upright; this slower pace can sometimes be imposed by an obstacle, difficulties of terrain, or a slope's gradient. It is a point that should be studied—and it would be a very practical study, in no way elaborate, but that we merely suggest should be carried out.

~

For shorter jaunts or longer excursions, Velocipedists naturally prefer flat routes to hilly ones, for while hills are great on the downhill slopes, they can be very hard on the way back up. Once the gradient is at a decimeter per meter, it is wise to dismount and walk alongside the Velocipede, steering by keeping one's hands on the handlebars. It is then a support, a kind of rolling cane, by means of which one can reach the top of a climb without unnecessarily tiring oneself out.

The descents are very pleasant when they are not too fast. One must always remain master of his Velocipede and keep control of the handlebar which also serves as a brake. When the slope is gentle, one can rest one's feet [on the flanges above the front wheel] and let the machine's momentum take over naturally; if the slope increases, there is a perceptible increase in speed, which may become considerable. In that case, it is necessary to put pressure on the brake and put one's feet back on the pedals, because just as the feet activate the movement, they can moderate it and even end it. Moreover, in advanced Velocipedes, the hand brake is very strong, almost instantaneous. The effort transferred onto the handlebar brake paralyzes the rear wheel, and forward momentum is almost immediately stopped.

This explains the need to safely control the handlebar, which is simultaneously the Velocipede's compass, brake, and pendulum. It acts directly, by an iron rod, on the front wheel where the pedals are affixed. It's the machine's soul.

When a rider is confident in his balance, he does not worry about the little obstacles that he is unable to avoid. While the Velocipedist dreams of macadamized pavement or asphalt sidewalks, he nevertheless rides on cobblestone roads without much concern for their bumps and irregularities.

~

At the end of a ride, if the Velocipede is too high to permit both feet to touch the ground at the same time, the best technique is to use the handlebar, leaning toward the center of the Velocipede's circular path and thus slowing down the movement. As soon as the Velocipede leans far enough, put one foot on the ground and immediately disengage the other one, while not letting go of the handlebar. The Velocipede's circular motion decreases almost instantaneously, and it comes to rest alongside the rider.

To turn to the right or to the left, one has simply to pull the handlebar toward oneself on the side in the direction one wants to move. It is important to avoid abrupt and jerky movements and to take into account the centrifugal force, a force that tends to push the rider outside of his anticipated trajectory.

It is therefore necessary, while steering with the handlebar, to lean slightly to the side toward which one wishes to go. There is no formal rule in this respect; the more rapid the speed and the smaller the radius of the curve, the more the inclination must be pronounced.

Experience alone will give a sense of how to position the body during a turn.

This explains why Velocipedists in the theater, who ride around in tight circles in a small space, have to lean in toward the center, sometimes managing to do so with a significant slant, making an angle of forty or fifty degrees with the ground.[58]

Beginners will do well to slow down and moderate their momentum when they want to change direction. The curved line is naturally favorable to maintaining balance, and it is less difficult to ride in a circle than to follow a straight line.

Expert Velocipedists prefer Velocipedes with large-diameter wheels to those that keep them too low to the ground. Racing speed increases proportionally to the wheel's radius with the same force exerted. But every coin has two sides, and it is all the more difficult to maintain one's balance the higher one sits. It is important to find one's happy medium.

It is a mistake to believe that it requires a great expenditure of force to make a Velocipede move. The calculations that we have given above reduce the effort required to about one-tenth of the fatigue of walking on flat, level ground; it is very little indeed. But, besides this effort, there are riders who unnecessarily exert nervous energy and tension in their muscles with no useful result. The movement of the legs must be supple, easy, and relaxed, and only ever in climbs should one feel a resistance that needs to be overcome.

When steered by a trained rider, the Velocipede can easily cover long distances and reach high speeds. Here is some information on this subject, taken from the most authentic sources.

In six days, two Velocipedists finished a ride of 150 leagues, the distance from Paris to Bordeaux, which equals an average of 100 kilometers per day. They did not overexert themselves and experienced only normal fatigue. Nothing would have prevented them from continuing their journey under the same conditions.

We also note—although it was set up as a challenge in response to a bet—a race of 250 kilometers, completed in twenty consecutive hours, including rest periods. The longest continuous ride that a strong Velocipedist can complete without stopping can hardly exceed 150 kilometers.

These measurements are what might be called travel speeds; they are far below the exceptional speeds that can be obtained in a short timeframe, by putting all of one's strength at the service of pride and ambition. It is possible to arrive at a speed of 500 meters per minute [over eighteen miles per hour], but this vertiginous pace cannot be maintained for long.

These exaggerations of speed are of interest only in sports; they cannot be used for serious study. It is similar for the eccentricities of certain skillful riders: letting go of the handlebar or posing as Fama or as an Amazon on their Velocipede.[59] We have even seen some go down the stairs of the Trocadero and ride on the parapets along the Seine: such games are bound to lead to a broken back.

Nevertheless, we cannot criticize Velocipede races without also casting blame on horse races; and that would make for too many adversaries. But let us suppose that, just as we try to improve horses, we want to improve the manufacture of popular machines. Velocipede races are certainly less dangerous for the riders themselves.

Moreover, our data on time and velocity, which the reader will find later in our article on "Velocipedes Races," are absolutely accurate, and we encourage curious readers to learn more about the results of human force applied to this type of locomotion.

The normal speed of the traveling Velocipedist is four to five leagues per hour, which is roughly triple the speed of a brisk walk. The movement that the body transfers to the machine should be as smooth as possible, using more flexibility than force since, as we have already stated, it requires very little exertion.

With this pace, it is possible to travel twenty leagues in five hours, without more fatigue than if one had walked for the same amount of time; there is even greater advantage in favor of the Velocipede since the forward movement is continuous, and the rider can take some brief moments of rest without stopping his momentum. This happens whenever going downhill or when going fast enough to let the machine coast on its own. He then takes his feet off the pedals and rests them on the nearby support, until he feels the need to add some speed.

In this way the Velocipede, of which we have already documented the utility, is not only fun but also a healthy, beneficial, and strengthening exercise.

As for the accusations against the Velocipede and complaints about it—that it weakens the nervous system or that it is the origin of maladies too intimate and unseemly to name—they are nothing more than old wives' tales. Exercise on horseback is ten times more dangerous from the point of view of health, not to mention the bucking, swerving, and bolting.

It would of course be impossible to deny that the abuse of the Velocipede has its inconveniences, but can't the same be said of all exercises, even of the mildest, the most pleasant, and the most comfortable in the world?

J. Legrand.

MARRIAGE ON A VELOCIPEDE

FIGURE 3.10. Mademoiselle Dorothy.

Let me tell you about Mademoiselle Dorothy.

She was truly pretty. I saw her every morning as I walked down the rue des Martyrs, in a glassed-in booth that her father had had installed in a corner of his shop.[60] She worked the cash register, receiving money from customers and returning change with a smile—and she lined up all the numbers on a big brass-buttoned till, a state-of-the-art machine that separated her from outsiders and made her seem invulnerable. Once money changed hands, she remained isolated from the rest of the world, and like a nun muttering her Our Fathers, she whispered numbers as the endless calculations entirely consumed her.

Less generous types took her for a wax doll. Woe to those who judge a book by its cover! I'm not one of those superficial people who are content with a glimpse, a reflection, an appearance. I go to the bottom of things, I persevere to get past obstacles, and if I find that a door is closed, I look for the secret formula that will open it. But what was the "Open, Sesame!" of this adorable person?

I had already spent a lot of money on knickknacks and other useless things like articulated dolls, tortoiseshell combs, mustache wax, hoops, and flying tops—here it might be helpful to know that her father ran a general bazaar—without having entered into the lady's heart. When I paid for my purchases, no more than a quick and discreet "thank you" would pass through her moist coral lips. I sighed like a forge bellows without her getting any more excited about my sighs than she did about the monotonous rumblings of the place Pigalle omnibus. Fortunately, the government came to my aid.

The government had resolved—and it can do what it wants—to start weakening French coins. Instead of striking them with nine-tenths of fine silver, as it had done before, it decided to replace some of the silver with excellent copper. This measure, which I cannot criticize, led to serious difficulties in the commercial relations of the population. Unified and with one voice, coachmen refused the older demonetized coins and declared that they would only accept money that was altered and thus reduced in value.[61]

This example of selflessness was soon followed by a throng of businessmen, and the young girl's father, who looked like he must have been in the National Guard, thought it his duty to lend a hand to the elected officials of France, specifically by strictly enforcing their decrees in his shop.

I learned all this by chance when I saw the attractive cashier scornfully pushing aside a ten-franc piece offered by a man of the cloth who had just bought a Pharaoh's snake.[62] After some discussion, the dignified man withdrew the offensive coin, which was adorned with the portrait of His Holiness, and replaced it with a skinnier effigy.[63]

It was a whole new horizon that I saw emerging, a horizon illuminated with the pink flames of hope. Pausing only to fill my pockets with prohibited coins, I returned to the bazaar with hopes of victory.

Her father was happy to see me. This amiable bourgeois, with the clever mind that comes naturally to traffickers, had noticed the negligence and distraction that dominated my purchases, and he took advantage of it to sell me balloons to decorate lanterns—that is, if the term *lantern* is not risqué.[64] Like magicians who get their victims to pick the right card, he plied me with old objects that had been cluttering his highest shelves. I didn't even haggle: such are the heroic acts that love justifies.

Yet, I shouldn't speak evil of him. That evening, the father had a good idea: "What are we going to sell to Monsieur?" he asked when I arrived. "A game of graces, a harmonica, or a Velocipede?[65] I've just received some newly forged ones from Michaux. They practically ride themselves, and you just need to give it a little push to get started."

This offer surprised me. I felt capable of walking out with a puppet, a rattle, or even a chariot drawn by his horse and loaded with parcels, but . . . a Velocipede! And so with some emotion that I could not suppress, I asked:

"What do you mean by a newly forged Velocipede?"

"I mean," he said courteously, "a Velocipede of recent manufacture, soft and elastic, both supple and sensitive, with velvet and steel in its springs."

"Very well, then. Wrap it up; I'll take it home."

And I went to the cash register. The beautiful girl did not move more than the Antiquities Museum's statue of Polymnia, whom she resembles a little in fact.[66] However, when the merchant's bellowing voice cried out, "A Michaux Velocipede! Make up a receipt for two hundred francs!" she seemed to come out of her silence. A quick glance enveloped me. I do not know if she recognized me, but a barely perceptible pink hue passed over her forehead: it resembled the fleeting and ideal complexion—noticed but for an instant—with which spring colors a wild rose's petals. I was momentarily disconcerted, but, regaining my composure, I put my money on the counter: a number of two-franc pieces bearing the effigy of Pope Pius IX, Leopold, Louis-Philippe, and other sovereigns banished from the Mint. Not a single authorized coin, acceptable head, or legal sovereign!

"Monsieur," she said, "that won't do!"

Her beautiful eyes, shining with indignation, suddenly rose. I had anticipated the attack; after all, I had gone through these steps only to get to this moment of supreme crisis. I endured her look bravely. I took it in and devoured its limpid rays. I wound it so tightly with my own gaze that she was unable to look away from me, and we both remained emotional and trembling, without really knowing where this magnetic energy would lead.

"Bah!" said the father with a conciliatory tone. "Let it go this once, if there is only one such coin!"

"But it's all of them!" she exclaimed, becoming animated. "Each and every one, from the first to the last!"

"No matter!" added the man. "We can have them exchanged at the Mint."

"No need," I said coldly. "I've changed my mind. Since my money is suspect, I won't take the Velocipede."

"What a joke!" said the merchant, troubled. "A customer like you won't take the Velocipede?!? Of course you will take it!"

"Yes, you will take it, sir!" said the girl.

"Miss!"

"You must," she said with a strange insistence. "A fashionable man cannot do without it."

"I'm not a fashionable man, Miss."

"And yet you certainly look like one," she said, obligingly.

"To prove the opposite," I replied, "allow me to tell you my story."

The father sat down, the young woman put her pen in her hair, and I began, breaking a deep silence with the story of my adventures:

STORY OF THE YOUNG MAN

FIGURE 3.11. The young man.

"I am not exactly of French origin. Although I was born in Montmartre, I was raised in Bougival, and Paul de Kock is my great-uncle.[67] The longer story of my family history is personal in nature, and that's a line I'm not interested in crossing. If it's all the same to you, let's not waste time on that, any more than on an inconsequential accident that befell my nanny."

"That's fine," said the merchant.

"Then I'll continue. I was destined for literature at the tenderest of ages, but, despite having received an excellent education, I failed to submit my application to become a member of the Society of Men of Letters, to which I have the honor of not belonging."[68]

"There's no need to humiliate anyone," said the father. "You can always resign on the 15th of August."[69]

"Don't interrupt him," said Dorothy.

"Indeed," I continued, "it saps all my resolve. Your charming daughter understands this, and it's all I can do to overcome the emotion I feel when in her presence."

"Does this mean that you like her?" asked the man gruffly.

I remained speechless in front of this honest man. Fortunately, a gentleman called him to haggle over a puppet. I looked at the girl. The doll had awakened; a crimson incarnate covered her delicate cheeks. I had my foot on the pedal.

"Ah! Dorothy!" I exclaimed.

"Hush," she said quickly. "My father's coming back! Take the Velocipede!"

I stood still, crushed by this command. Just when I thought I was nearing her heart and touching a chord, a lowly mercantile preoccupation still dominated her! My illusions were crushed: I was like a man who thinks he'll reach out and grab a star and who, feeling drops of oil on his nose, discovers that he's only holding a kerosene lamp.

I was pulled away from these thoughts by the father's voice, which suddenly seemed much warmer.

One cannot handle puppets without consequences. Haggling with the client over a puppet must have reminded him of the highs and lows of existence. Perhaps he foresaw a future for his darling daughter; perhaps the silhouette of this silly plaything reminded him that she was old enough to be cared for. Whatever the reason, in approaching us he was seized with a strange tenderness.

"Young man," he said to me, "I was once twenty years old like you. Believe it or not, I was not born to run a bazaar in the rue des Martyrs. The trust you place in me obliges me to speak. You will know my story, and in it you will find useful lessons. And you, my daughter whom I adore, forgive me if I have so long hidden my secrets! The time has come for you to learn everything."

Her eyes bathed in tears, Dorothy came out of her nook and threw herself into her father's arms. The latter, after blowing his nose, began in these terms:

STORY OF THE MERCHANT OF THE RUE DES MARTYRS

"I was born in Mantes-la-Jolie, which is proud to have given birth to Charles Monselet and Anténor Joly.[70] It is a pleasant city, with lively fountains and sand-blasted walks, of which its municipal council takes great care. The women are graceful and welcoming; the men are just as polite in spite of their inveterate habit of wearing spectacles."

"I imagine," I said to the merchant, "that they don't all wear glasses. It's not as if simply being in Mantes—"

"I was wrong to believe it once," said the old man with a heavy sigh, "and that was the source of all my misfortunes. Raised by an even-handed father, I wore glasses like the others, although it bothered me quite a bit when I tried to sleep, and when family affairs forced me to leave for Switzerland—"

"Oh, Father!" cried Dorothy. "What are you about to say? Sir, do not let him continue!"

"Why not?"

"Don't you notice his eyes wandering and his voice trembling? Alas! He has undoubtedly left terrible secrets behind in Switzerland! It was there that he met my mother. She had wanted to hike along the glaciers and the mountains with him. 'No,' he said. 'I know that you are imprudent; you will want to go everywhere, you will not listen to the guide, your foot will slip . . . and so you will not come!' 'I say that I shall go!' 'And I say that you shall not! What pleasure would travel bring me if I saw you doing stupid things?' 'Accidents never happen.' 'Why yes they do!' 'Why no they don't!' And she was stamping her feet. 'Well then, I will go with Ferdinand!' He was a slacker, a friend of my father, whose laid-back nature gave my father balance and perspective, even though he could be stubborn when he wanted to be."

"Let me go on," said the father.

"No," yelled the child. "When disaster strikes, when Mother falls in the ravine, you will make a terrible scene. Sir, in the name of Heaven!"

"Well, then!" I said, "Let's leave it at that. It's best I just take the Velocipede."

"Very well," said the merchant. "And since you have no suitable money, I will have it brought to your home."

I bid them farewell, very perplexed, exchanging with Dorothy a look that went to the depths of my soul. The Velocipede was delivered the next day, and it was very difficult to bring it up to my sixth-floor apartment. It made it, in the end. I spent the day practicing on the balcony in front of my windows, going forward and then backward, so that at ten o'clock in the evening I knew what I was doing.

The next day, quite proud of all I had learned, I asked the local handyman to carry my Velocipede on his back, and I headed toward the rue des Martyrs, with the intention of prancing around and performing some fancy dressage moves in front of Dorothy's store.[71] Imagine my surprise when I cast my eyes on the glassed-in booth where she usually sat and saw no one! The niche was deserted, the ledgers abandoned; her father was standing on the sidewalk, smoking his pipe, alone and carefree.

"Where's Dorothy?" I asked him.

"She's in her class," he said grandly, "at Michaux's riding school."

"Which class?"

"Her Velocipede class."

"Heavens!"

I understood everything in an instant. I ran to the gymnasium in great haste but a concierge, immovable as destiny itself, forbade my entry, under the pretext that it was the ladies' class. I tried to bribe him; he was inexorable, even when I offered him some crowned coins. Nevertheless, his strictness softened somewhat.

"You cannot come in," he said, "but nothing prevents you from waiting for these ladies at the exit. In fact, today is the day they dive into the deep end, when they leave the school to try riding outdoors. Our Amazons will leave for the Bois de Boulogne. In fact, here comes the cavalcade now."

I shuddered in admiration. Like Calypso in the midst of her nymphs, Dorothy, calm on a fiery Velocipede, came out at full speed, kicking up dust behind her. Twenty young beauties formed a procession around her. The elegant squad had just reached the Champs-Elysées when it shot off like lightning. There were some pretty legs there.

I saw only Dorothy's. Oh! The leg of the woman we love! Does it need to be said that I, too, had climbed up on my hobbyhorse and that I was flying in her wake?[72] In a quarter of an hour we reached the Arc de Triomphe de l'Etoile; five minutes later, we entered the Bois de Boulogne. When the beautiful Dorothy stopped for a moment, I took full advantage: riding in a small circle, passing right before her eyes, greeting her with exaggeration.

She uttered a cry, like that of a frightened bird, and, like Galatea, fled off down a side path with twists and turns that disappeared into dark shadows.[73] Her Velocipede's wheels turned so quickly that their spokes were no longer visible; her little boots, agitated by a feverish movement, stomped down with all her might, quickening her speed; flying behind her in the wind was her beautiful light-brown hair, with golden reflections, like the luminous tail of a comet.

There wasn't a moment to lose. I intuited the secret that Dorothy hid in her pretty, thoughtful head. My happiness went straight to my legs: I raced after the young woman, knocking over a sprinkler that was in my way.

I was not just riding: I was flying with the velocity of an arrow. A kid passing by shouted, "Look out, here comes the express train!" Dorothy, a hundred yards in front of me, kept her distance at first. In terms of her vivacity, she was every bit as quick as me; all I could hope for was fatigue. She turned off the pathway onto the broad Longchamps promenade and headed toward the waterfall.

FIGURE 3.12. Love on velocipedes.

A feeling of intense joy came over my soul: I was noticeably gaining on her.

At the lawns in front of the waterfall, she was only sixty paces ahead of me; she took a decisive, abrupt left turn and entered the Vierge-aux-Berceaux path.[74] Five minutes later I caught up to her, and the curls of her hair brushed up against my face.

"Dorothy," I said, "I love you."

She did not answer. I made one last effort and went one head length ahead of her.

"Well!" she said. "I will love you, but you would never have known, if you had not passed me."

"What do you mean, dear child?"

"Ah!" she said with a bit of melancholy. "It's quite a story. My friend, slow down and let's take this quiet alley. You will learn why you had to win my hand by the strength of your legs."

"I'm all ears," I replied.

She took a moment to catch her breath, for the ride had made her breathless; then, in a sympathetic and melodious voice, she began her story:

STORY OF THE DEMOISELLE WHO COULD MARRY ONLY THE MONSIEUR WHO COULD BEAT HER AT A RACE

"My mother," said the charming Dorothy, "was a shepherdess from the canton of Appenzell, renowned for its cheeses. You no doubt know that Switzerland owes its deliverance to William Tell?"

"Indeed," I replied. "They made an opera about it."[75]

"As a descendant of his, on my mother's side, I have inherited a very independent nature. I was born in that country with its open air and mountain peaks, where you breathe freedom fully into your chest. I helped my mother tend her flocks by day and in the evening, while the *Ranz des vaches* played, I enjoyed watching the cows parade by, row after row."[76]

Saying these words, the young enthusiast's bodice rose, throbbing like a dove caught in a trap, and I dreamed of spheres and snowy mountain peaks.

"Oh!" she said. "The Alps and their wild grandeur! Torrents and abysses, wide-open spaces, and lost horizons, the immensity that

FIGURE 3.13. Angel carrying velocipedes.

makes you shout: Give me wings! That's what cradled me in my youth and what I missed in Paris! I was suffocating in its narrow streets, in its squares, in its crossroads. And do you know what saved me? The Velocipede! The Velocipede that gave me back energy, wings, the wind blowing in my hair, the voluptuousness of speed!"

"And so this," I replied, "is why you are no longer silent."

Eight days later, I married Dorothy at Notre-Dame de Lorette. We spent our honeymoon riding in Buttes-Chaumont park.[77]

L. G. JACQUES.

ORIGIN OF THE VELOCIPEDE

FIGURE 3.14. Letter to Madame.

To: Madame[78]
Madame *La Librairie du Petit Journal*
Paris

Dear Madame,

You ask me about the Velocipede's origins: not from a prosaic or vulgar point of view but from that of logic and learned research. It would not be difficult for me to write thick tomes on such a subject, and I would not refuse to undertake such a task, if you were to find me a publisher open to the idea.

However, this is not the time to freely show off all my knowledge and, for the pleasure of your readers, I will restrain myself and stay within the limits that you have set.

In metaphysics, it is commonly understood that the ideas of man, as they are related to inventions and material imagination, derive from the observation of nature rather than from any illumination or intuition. This amounts to saying that man does not create; he imitates, applies, and perfects. It is impossible to draw a line or a shape whose natural prototype has not already been seen. No conquest escapes this somewhat humiliating rule: lightning and magnets have taught us electricity and magnetism; beef stew is the father of steam.

I am not seeking to diminish the merit of the elite minds who keep us moving along the path of progress. But their ability to create does not stem from their originality. For centuries, small birds have been solving the problem of aviation before our very eyes; the one who brings human flight to the world will therefore be only a copyist: we should not despise him for that.

The horse—conquered by man from the earliest times—is certainly the guiding idea, the germ, the basic type for a whole series of devices of locomotion that have enriched humanity. In this series, we will distinguish a special class that contains the machines intended to imitate the horse or to replace it: machines moved by human force alone. This definition opens up a whole range that begins with a stick and ends with the Velocipede.[79]

Yes, quite seriously, the first child riding a stick like a horse is the primitive inventor of the Velocipede. Everything else is just improvement after improvement. It is even possible that the child in question was Abel, or his brother Cain, which places the origin of this machine at the creation of the world.

The hobbyhorse should not be relegated to childhood games. Even today, Alpine shepherds descend the snowy slopes astride their long-handled ice axes; there is no tourist who has not experienced this kind of locomotion.

If we move ahead to the Great Flood, we can easily find the Velocipede in Noah's ark.

Scripture, which delights in parables, speaks of an olive branch brought back by a dove. Our contraption derives from the stick just as the stick comes from the branch. To descend the slopes of Mount Ararat, slippery from forty days of rain, our forefathers had to arm themselves with herdsmen's sticks.

Think back to the ancient times that we have just sketched out. In this curved stick, do you not see the head of the horse and its proud gait? The Velocipede's progress is slow and steady. In early civilizations, it first imitated a horse's form and contour; and later, it would set out to conquer speed.

Let us put aside this first period, from which emerged two significant and distinct currents, each rich in analogies. On the one hand, the stick leads, directs, and governs: it is the Catholic cross in the hands of bishops and abbots. On the other, it leads astray, misguides, and leads naked witches to the Sabbath: it is the broomstick.

If from theology we move forward to autocracy, the mechanical horse takes shape in a precise way; its form is definitive from the earliest days. Its most famous personification is undoubtedly "this famous horse concealing heroic Greeks in its flanks, and carrying into frightened Troy desolation, carnage and death."[80]

If this epic steed leaves nothing to be desired in terms of tragedy and the picturesque, it seems to have been difficult to handle, since the Greeks could manage to push it into Troy only with some help from their enemies.

Let us now skip ahead a few centuries to arrive . . .

In the happy age of chivalry.[81]

In the Middle Ages, the legend of the Velocipede or the Wooden Horse is everywhere. There are few stories that do not include it. Cardonne and Petit de la Croix found theirs in Persia; Galland makes one appear in the *Thousand and One Nights*.[82] He writes:

> An Indian presented himself at the foot of the throne; he brought with him a wooden horse, saddled, bridled and richly harnessed, decorated with so much artistry that to see it, one would have taken it for a real horse.
>
> As soon as the king had made his will known, the Indian turned his ankle ever so slightly, turning it in toward the horse's neck and raising it to the saddle pommel. In an instant, the horse leapt up from the ground and the rider was off, fast as lightning.

If I weren't wary of moving too far from my subject, I would tell you the story of Astolfo's hippogriff, but Ariosto is too fanciful a storyteller to appear in a scholarly study. I'll leave the hippogriff to the fairy tales.[83]

I do not have the same concerns about the famous horse named Chevillard, mentioned in *The History of Magalona, Daughter of the King of Naples*, a precious little volume printed in Seville in 1533. The author writes:

> This wooden horse is steered by means of a peg that he has in his forehead and that serves as a bridle. He flies through the air with such speed that it looks like the Devil is carrying him. According to an ancient tradition, he was manufactured by Merlin the Wise, who lent it to his friend Count Pierre, who loved the young Magalona. The count stole off with the princess, put her behind him on the horse, and took to the air, amazing all those who saw him pass. . . . What is good is that this horse does not eat, does not sleep, does not wear through horseshoes, and strolls easily, without muscles or nerves, such that whoever rides him can hold a full glass of water without spilling a single drop, he strides so steadily and calmly. That is why the pretty Magalona was so delighted to be riding on his back.[84]

It is difficult to dispute the testimony of Cervantes, from whom we borrow this description. It may be objected that he makes Chevillard play a ridiculous role, by pretending to have him carry Don Quixote and

Sancho Panza off into space as they are fanned with forge bellows and smoked like hams.[85] But Cervantes wrote a parody and never pretended that his Chevillard was the real one.

In Louis Viardot's notes to his edition of *Don Quixote*, we read that old Chaucer, the Ennius of English poets who died in 1400, spoke of an almost similar horse, which belonged to Cambuscan, king of Tartary. Made of bronze, it was directed by means of an ankle placed in its ear.[86]

If we take into account the colorful style of the time, it seems that we are not far from the modern Velocipede. It is not yet fully formed, and this next story may take us beyond it. In one of the funniest books written for children in recent years, we read *The Story of King Mistanflûte*, a traveler whose adventures exceed by a hundred cubits the most incredible travel narratives.*[87]

We will borrow only the episode from his odyssey that relates to our subject.

King Mistanflûte, passing through Lisbon, dines with an inventor who has just invented a new kind of mount. He obligingly describes it as follows:

> Imagine a horse's body, mounted on very light wheels, which can be powered by the feet or the hands. You have seen them on the boulevards in all the big cities; there are children's toys based on the same idea and which toddlers operate in living rooms or in gardens. This is nothing new. Here is where we deviate from the beaten track:
>
> In the horse's body, which is life-sized, are very powerful steel springs, which add force and, without becoming too intense, keep it going for as long as possible. This force is similar to that which is obtained by winding a stopwatch but exponentially more significant. When we bend or compress a fixed metal bar, due to the laws of elasticity, it tends to return to its original shape and

* *The Story of King Mistanflûte* is part of a four-volume work, as interesting as it is instructive, published by the Librairie du Petit Journal under the general title *Le Château de Robert Mon Oncle* [My Uncle Robert's Castle]—for the price of six francs.

position. This produces momentum, which makes a watch run for fifteen days, a month, or more.

Furthermore, this kind of driving force is very well represented today in toys with which we are familiar. I am thinking about mice or little cats, mounted on caster wheels, filled with inner cogs and sprockets, that we wind up like pendulums and let run around at will.

The Portuguese horse was a simple rolling mouse but made on a large scale, with particular improvements. It was necessary to exert a persistent force to wind the springs with enough tension, and this preparation took nearly an hour. The horse could then walk for about two hours on a flat surface, covering a distance of fifty kilometers. But here is where the beauty of the invention lay particularly . . .

The rider could suspend the springs' action to manage the loss of force. It would then stop naturally, if the road was flat, as soon as it had used up its momentum and speed. But, as long as it was on a downward plane, its distance would be extended more or less, according to the incline. If the slope was very fast, and he arrived at a high speed, he would be obliged to use a special brake within his reach. However, the action of the brake was ingenious in that it wound up the springs and stored up new energy. This is too obvious for me to explain it at length. The brake in this case was the hallmark of ingenious engineering.

The inventor nonchalantly explained that, having left in the morning from Santarém, he had arrived in Lisbon, after a journey of a hundred kilometers, without having tired his horse. It was a marvel; but the route traveled generally sloped in the direction of his destination, thereby explaining his exploit.

After the meal, we went down to see the curious machine. The ladies wanted to ride it and try it out. The inventor explained the ingenious arrangement of the wheels, which were made of two iron hoops separated by a number of little angled flaps, roughly resembling the blades on a water wheel. He had foreseen the possibility of crossing a river, and this natural obstacle would not stop his machine from working. The horse floated without difficulty, and the waterline did not reach the openings through which the transmission of movement took place. Its wheels, three-quarters submerged, beat through the water in which they were plunged like a steamship. All of these details filled us with enthusiasm, and in his admiration for the spring horse, one of the guests exclaimed, "It only needs wings!"

We took a tour of the park, and to satisfy our curiosity for trying out more serious tests, the inventor entrusted the machine to his servants, who wound it back up.

"It is sometimes a little lively at the start," he said laughing. "Triggering the springs gives quite a jolt. Who wants to try it first? Will it be you, Mr. Frenchman?"

Despite my ignorance of the expression he used to refer to me, I understood his offer very well, which he accompanied with a mocking smile.

"With pleasure," I replied, stepping into the stirrup.

As soon as I was in the saddle, he stepped forward to give me some useful instructions. But, with a poise that grew out of some bizarre self-esteem, I leaned on the lever that started the machine, and I sped off like lightning, leaving everyone stunned and the inventor a little worried.

My imprudence, however, was not as great as one might suppose. I had been studying its mechanics while the ladies were trying it out, and I found myself rather skillful. The horse's pace was the smoothest in the world, and except for its extraordinary speed, I was no more uncomfortable than if I were in a carriage. The axis, which allowed me to turn the beast to the right or to the left, pivoted perfectly, and I rode around in large circles, skillfully avoiding all obstacles.

I heard the sound of applause, which intoxicated me somewhat. I steered my path toward the group of spectators, and to thank them for their cheers, I raised my hat with perfect ease. To see the ladies' smiles a bit longer, I turned my head for a split second; that's where things went wrong.

The front wheel, which I misdirected, violently struck a post that was flush with the ground. I was not taken aback, but I felt a rude shake in the hand that was holding the steering lever. The steel rod had broken; the horse no longer obeyed. At that moment, I was launched at full speed down a wide alley of orange trees that ended at the beach and which had as its horizon only the distant blue of the sea.

A cold sweat beaded on my forehead; I quickly considered my few slim chances of safety, and none seemed plausible to me. I thought of jumping to the ground, but I would have shattered like glass, as the speed of my mount was increasing with each moment. The road tilted toward the sea; I passed like a swallow under the thick arches of branches. Suddenly sunlight swept over me: I arrived at the shore. I hoped for a moment that the sands of the shoreline would slow my progress, but the waters were too high for that. Wet dust kicked up when I entered the water; water went over the blades on the wheels. I felt that the horse was losing its footing, and I was lifted limply by an immense wave that carried me away. The wheels violently whipped the waves they were crisscrossing as I hurtled on toward the open sea, toward the solitudes of the Ocean! I tried to turn around, to change direction, shaking my legs sharply. It was all for naught, as I had a tailwind.

The sun was setting. Some boats passed nearby; I hailed them with all my strength. I distinctly saw their crews frantically make the sign of the cross and then put all their force into their oars to get away from me. I cursed at them, which I now regret; it would have required unusual and singular courage for them to approach the sea monster I was riding and to face the whirlwinds of foam that I was whipping up around me.

I will stop here, my dear friend, because I do not believe I have much more to say. I have not only delved into the past, I have shown you the future. You can see that the Velocipede is grounded in tradition and has its very own pedigree and nobility. Who knows where the genius of man can make it go?

Yours,

De la Fredière,
Of the Institute.

VELOCIPEDES OF THE PRÉ CATELAN

FIGURE 3.15. At the Pré Catelan.

The intelligent and witty director of the Pré Catelan, Mr. T. de Saint-Félix, on the lookout for all the new products and innovations that can excite public interest or curiosity, has given Velocipedes a central role in the field's annual schedule of spring and summer festivals.[88]

He understood the practical value of this new kind of locomotion and the improvements it could bring to social relationships for inhabitants of cities and of the countryside alike. To this end, the administration of this marvelous park, so aptly called the Flower Basket of the Bois de Boulogne, is organizing, this summer, a number of Velocipede *Competitions*, *Races*, and *Exhibitions*.[89]

The *Competitions* are intended to facilitate comparative studies that will make it possible to develop this branch of industry and to adopt the most advantageous methods of manufacturing. The popularity of Velocipedes is increasing every day, and this kind of friendly competition offers precious encouragement to this growing industry.

The *Races* allow spectators to see individual progress in the art of cycling. It is not only a question of being the fastest but also of riding safely while maintaining perfect balance. Hence different kinds of activities: speed trials, obstacle races, and slow races. It should not be forgotten that although the Velocipede is fun and a luxury for many people, it has its useful and practical side, which is such an important part of its promising future.

Finally, the *Exhibitions* are intended to show off the Velocipede's new models and new uses, which are far more numerous than we think.

It is difficult to find a better place to meet than the Pré Catelan for such serious affairs. With its smooth and sandy paths and gentle slopes, this garden has for many years been the meeting place for the elegant regulars of the Bois de Boulogne and Parisian high society.

From spring to autumn 1869, there will be national and international festivals, and Velocipedes will be front and center. There is even talk of ladies' races, and it would be unfortunate if Monsieur de Saint-Félix confined them to the Hippodrome, because they are very attractive and would in no way shock the public, if they are properly managed.[90] A jury, made up of experts, will judge the competitors and award the prizes. The most scrupulous impartiality will preside over its deliberations.

There will therefore be a Veloce Club in Paris, with its headquarters located at the Pré Catelan, which will oversee these exhibitions.[91]

The Pré Catelan will open on Easter Sunday, and the first Velocipede races will take place the next day, Monday. We do not need to remind readers of the appeal of its musical matinees, its children's balls, or of the Theater of Flowers, where several original plays will be performed this year. But we wouldn't want to advertise.

RAYMOND.

FIGURE 3.16. A conversation.

POLITICAL STUDY OF VELOCIPEDES

FIGURE 3.17. Riding in the rain.

Eight hundred and seventy-five years after Idomeneus founded Salento, when the Cretans had shaken off the yoke of the Turks and adopted the Jacotot method in their primary schools, there came a time when the state's finances were in very sorry shape.[92] This disorder could not be attributed to ministers' wastefulness or to the opening of railway lines, because the ministers were paid only two francs an hour and the railways hadn't been invented yet.[93] Nevertheless, significant financial need paralyzed commerce and banking; the stock market had fallen so low that investment income was below all estimates, and stock certificates were sold to hairdressers who used them for curling hair. This situation dragged on for so long that the Emperor Mouchette grew increasingly alarmed.[94] This esteemed prince adored his people, especially from the point of view of the income he received from them. His naturally fatherly heart was interested in their happiness, even though he gunned them down in the streets at the slightest hint of a public disturbance. Many people were rotting in exile and in dungeons for having expressed feelings of loyalty that the prince mistook for bitter personal criticism. He loved his homeland for what it brought him.

If ill-advised readers saw any specific criticism in the preceding description, they would be greatly mistaken. Emperor Mouchette was neither better nor worse than most of the potentates of his time. Since then, things have changed considerably.

The Cretan emperor had just finished lunch and was cleaning his dentures with a small brush—because he was still a kind of dandy, even though he had fallen on hard times—when his valet entered the dining room with an alarmed look. He told the monarch that his grand vizier, Poulette-Matapan, was requesting an immediate audience.

"Let him come in!" said the sovereign, who wasn't experiencing his usual gastric distress that day.[95] "He can have coffee with me."

But a quick glance at his guest's face made it clear that coffee would not interest him. Poulette was distressed, and his esteemed host felt all the weight of his silent devastation.

The emperor knew that his vizier was not easily unsettled. His affection for him was both great—from when the vizier had taught him the recipe for his snipe stew with turnip jelly—and abundant—from when the vizier had loaned him money when the imperial coffers had run dry. Historical accuracy forces me to say here that the august Mouchette was more than a little gourmand. He had once traveled ten leagues on his favorite mount just to pick fresh morels. He had returned with indigestion, while his steed, which he had overworked, had fluid in its lungs. They had tried to treat him with iodine and geranium infusions, but he had perished within twenty-four hours. Here it should be understood that I'm talking about the animal, not the emperor.

The emperor had wept for this faithful beast. But even the greatest sadness must come to an end, and the vizier, hoping to distract his master, replaced the deceased animal with a Himalayan giraffe: perfectly trained, able to trot perfectly and bark like a seal.

Since we become attached to people whose good intentions appear as delicate attentiveness, the sovereign was naturally very fond of the vizier, who returned his affection in kind. That's not to say that Poulette would have been the last to insult him or cover him with mud if his people succeeded in chucking him out, but, until that day came, he promised his unshakeable loyalty and attachment.

The Emperor Mouchette therefore counted on the minister's devotion in the right way.

"Well then!" he said, when he saw the vizier arrive with a distraught appearance. "Am I to assume that there are protests in the faubourgs?"

"If only," said Poulette. "That would be cleared up with a bit of gunfire and a court-martial. This event is of the utmost gravity."

Upon hearing these words, the prince stared at him from head to toe; it took him a moment to fully take in the strangeness of his appearance.

Poulette was dressed as Pierrot, and as Carnival was over, it seemed more than a little unusual.[96]

"What is this ridiculous costume supposed to mean?"

"You will understand it all soon enough," said the vizier with emotion. "My duty obligates me to show you the abyss that is opening up under your feet. You know that the heavens, to our dismay, have given us disturbing neighbors. We are separated from the Liffre-Loffres only by a small stream, which is currently dry.[97] Our enemies, under the pretext of looking for crayfish under stones, have crossed the Rubicon and covered the walls of your palace with indecent effigies and graffiti."

"What!" said the emperor. "Without checking with the censors first?"

"That's correct."

"Did they at least request permission?"

"They didn't even do that. Still, it would be nothing had they stopped after writing these immoralities. But would you believe that they also dared to compare Your Majesty's stately face to a cooked apple?"[98]

"There are no limits anymore," said the monarch.

"It sets a bad precedent for your subjects. War therefore seems imminent to me."

"Certainly," said Mouchette. "Especially if I am personally attacked."

"Now," added Matapan, "you have to face it head on; for if we start a little quarrel with the Liffre-Loffres, there's no doubt that they will beat us up."

"No kidding!"

"They'll steamroll us flat, I maintain, unless we stick some spokes in their wheels."

"Wait, so they have wheels?" said the dazed emperor.

"Therein lies the mystery," said the vizier in a low voice. "You should know that this is a state secret and that you must not gossip about it with the empress or the little Chiffonnette."

"My lips are sealed."

"I hope you're right. Did you notice that the king of the Liffre-Loffres had his garden covered last year with a huge canopy under the pretext that his shrubs feared the great outdoors?"

"Of course, since there was even a diplomatic statement about it."

"Well, sire, we were being had. The clever monarch simply wanted to conceal his equestrian center."

"An equestrian center!"

"A cavalry equestrian training ground; and here's where you can probe the depths of your rival's politics. He saw, as we did, our army's natural disadvantage stemming from our lack of draft and saddle horses. These animals, whose utility cannot be disputed, especially for horse-drawn carriages, cannot live on our island. They are struck with a malignant pertussis within eight days of their arrival and die right away, despite the licorice root that they are fed exclusively.[99] In sum: no horses, no cavalry!"

"Do you really think so?" said the prince.

"Certainly. And what has happened? While you were handing out prizes to your academies to combat the scourge, our enemy had a stroke of genius."

"He had a stroke?" said Mouchette, gasping.

"Oh, it's very ingenious: if you can't cure horse pertussis, you have to have horses that don't come down with pertussis."

"But therein lies the difficulty!"

"Our rival figured it out. In Paris—an industrious city located under the 49th parallel—he had horses made of wood and iron; they are impervious to the infirmities of nature and to bad weather."

"And what do these extraordinary creatures eat?"

"They don't eat anything. A kind of Prometheus makes them in workshops where there are more noisy hammers than ancient Lemnos ever heard.[100] The army of Liffre-Loffres, mounted on these tireless steeds, comes and goes, flies, multiplies, and is ready to hit us with all its might."

"The situation seems tense," said the emperor. "What if I went down to the cellar? I have some very good wine that I need to put in bottles right away."

"I wouldn't do that. You would demoralize your soldiers. On the contrary: show yourself. This morning's demonstrations are insults that you must not accept quietly. Maintain your composure, Your Majesty; I'm going to work in the shadows."

"Ah! You are going to work in the shadows?" asked Mouchette. "What if I accompanied you?"

But the vizier was gone.

~

Alone at his table, the emperor had a large glass of Spanish wine brought to him. He seemed to hear distant sounds; his legs shook, his ears were ringing, and worry clouded his eyes. However, he knew the duties of his birthright, and after half an hour of hesitation, he called out loudly, like a man who was ready to act.

"You, there," he said to the first servant who appeared, "go prepare a coach and have it waiting for me at the side door. We'll leave on the hour."[101]

"Sire," said the surprised servant with a little embarrassment, "there are significant riots in the city; the people are organized and chanting. Your Majesty would perhaps do well to intervene."

"That's why I'm going," said Mouchette.

But, as he stepped over the armchairs to move even faster, a great noise was heard; the gates of the palace cracked, yielded, and gave way to the swarming crowd.

"It's too late!" said the emperor, turning pale.

But he suddenly recovered upon seeing his courtiers adorned with feathers, the empress and the vizier leading them, crying out, "Victory!"

"What!" he said. "What is it? What's going on? What happened?"

"What happened," said Poulette, still breathless, "is that I met our enemy's caretaker, and I tried to corrupt him. He objected, citing his oath of fidelity; I added twenty more francs. The machines are ours; I've hidden them in the cellar."

"Triple ox horn!" cried the emperor. "Kneel down, so that I may give you my blessing!"

"As you please!"

"Do you want the medals of my orders?"

"It's all the same to me."

"The Golden Fleece? The Garter?"

"Go right ahead."

"Here, kiss Eulalie."

Poulette-Matapan threw himself into the arms of Her Majesty, who was getting on in years, but was still very easy on the eye.

"Everything is for the best," said the monarch, out of breath.[102] "But I still don't see why you dressed up as Pierrot."

"Neither do I," said the minister. "At least it didn't cause any harm. Let's ride, gentlemen, let's ride!"

In the blink of an eye, enthusiasm rolled throughout the courtyard, tracing a thousand arabesques on the pavement in the main square. The king of the Liffre-Loffres, high up in his attic, watched the celebratory commotion of what was, for him, a day of suffering. With his characteristic diplomatic skill, he immediately put on a coat and came to compliment the emperor of the Cretans, while communications officials raced to and fro, erasing that morning's caricatures.

Thus, a minister's wisdom prevented a bloody war between two great peoples. In addition, honor redounded to mechanical horses; in memory of this illustrious day, they were baptized with the name of Velocipedes.

Doctor VABONTRAIN.[103]

FIGURE 3.18. Velocipedist memorialized.

ON THE TOPIC OF EXERCISE

FIGURE 3.19. Taking a break.

Physical exercise, just like proper medical treatment, needs to be individualized, that is to say, each situation requires principles that are appropriate with respect to age, sex, and temperament.[104]

Adolescence is the time when physical exercise is the most useful; it helps educate the senses and the musculoskeletal system.

During puberty, it has the effect of distributing the exuberant vigor that tends to concentrate in the sexual organs throughout all the muscles, and to prevent the vicious habits that these organs' excessive sensitivity too often determines. Neither morality, threats, punishments, nor shackles can counter these dire tendencies. The only ways to prevent or destroy them are through physical fatigue and rigorous muscular exercise.

In adulthood, exercise is still useful in order to maintain balance between all the parts of the organism and to avoid the vital concentrations that could build up in the organs; it is especially so for people whose occupations are sedentary: for men of letters, scientists, and office workers.

Finally, moderate exercise is also suitable for the elderly; it helps the organs run smoothly and requires the use of fibers whose sensitivity is otherwise dulled by inactivity.

As you can see, the benefits of exercise are numerous; but its abuses are no less so, and we think it advisable to roughly sketch out the different kinds of exercise that we think should be adopted, according to the various conditions indicated above.

The child is a soft wax that can be stretched in any direction; although, if the exercise is too strenuous, the goal would be exceeded, with disastrous results. Therefore, one must focus on flexibility and physical development, sensibly choosing the right methods. During childhood—that is to say until the age of ten or twelve—one simply doesn't give the slightest thought to what constitutes physical labor; the child's growth could be altered by it, as could his or her character. It is important to do the necessary amount: neither too much nor too little.

The woman, who is a big and admirable child, needs the same consideration and the same delicate attention as the child does. Like the fragile shrub that resists the hurricane better than the old oak tree, she needs to preserve all the forms and graces unique to her sex so that she can acquire the energy she will someday need to conceive and give birth without danger. Women need supple movements, gentle bends that make their limbs flexible, developing their breasts and strengthening their core. Let us create, for lack of a better word, a relatively strong being—within the limits of what is possible, and suitable for her own needs—but let us be careful and avoid creating a female Hercules.

Middle-aged men should not exercise with swings, the fixed bar, the pommel horse, jumping pit, etc., and they should engage almost exclusively (unless we are dealing with an exceptionally slender and vigorous subject) in general calisthenics or in well-thought-out and gradual movements, including the use of exercise equipment.

So, except for young men between fifteen and thirty years old, nothing violent, nothing excessive, graduated efforts, instruments and weights always lower than the acquired strength, and with the predominant goal being balance and health.

Important notice:

The time of day when one exercises is not unimportant.

Based on our experience, we can offer the following advice:

Mature people should take morning classes, before second breakfast, or evening classes, before dinner.

People who are overweight, those who suffer from gout or diabetes, and those with high blood pressure should take morning classes, preferably on an empty stomach.

For people who are nervous, bilious, lymphatic, or anemic: take a class that precedes dinner, adding to it a shower or a wet towel rub.

In general, leave an hour between the end of an exercise session and the next meal, and wait to begin exercising until at least three hours after the last meal.

Only healthy and vigorous young people should get in the habit of exercising at all hours of the day, in order to be always ready to exert themselves and to be able to successfully endure whatever tests the future may hold for them.

What is a shower? Readers will ask us.

Water at a specified cool temperature, projected either perpendicularly, horizontally, or upwardly onto the body. It is a series of vigorous, almost violent affusions: watering, immersing, spraying the epidermis in all directions. It puts the blood in a temporarily abnormal state of circulation, driven back first from the surface to the center and then returned with a new force toward the periphery of the body. The nerves are startled, energized, awakened from their torpor by the shock of the jet stream and the sensation of cold; the skin's pores contract and open out, breathe in large quantities, get rid of everything unhealthy they contain, fully appropriating, we can say it here, a whole new vitality. In a word, the whole bodily economy is ramped up, producing an effort that elevates the body's strength and returns the acts of assimilation, secretion, and excretion back into the range of normal bodily function.

For the shower to produce the desired effect, it must be both an action and a reaction.[105]

Let us explain.

Whoever takes one must, whenever possible, be in a state of perspiration: not the artificial perspiration of the steam bath or that which can be brought on instantaneously but specifically the excellent sweat that sustained effort from well-thought-out and organized work brings to the surface of the skin. Fencing, boxing, horseback riding, or any other physical work—all of it is good; in our opinion, what seems preferable to everything else is exercise itself, for with it all the limbs are brought into play one after the other, all the organs are subjected to an equal degree of exertion, all the tissues open up and are ready to receive the striking impressions of opposites: cold opposed to hot. The feeling must be terrifying for its consequences to be satisfactory; the whole keyboard of the human mechanism must give off a sign of life in a supreme thrill.[106] When this is done, action ceases and reaction begins.

Suddenly receiving a large amount of water on the body for a minute or two is obviously, as soon as the shock has passed, a way to cool down.

This should be followed immediately by dry friction, which should not be confused with the friction from a glove or from a towel soaked in cold water.

Dry friction, to be clear, should not simply be letting the wet epidermis air-dry; it must, before and above all, be an immediate and brutal drying action like the very action whose excessive energy it is meant to temper.

Therefore, you must not, for reasons of comfort, avoid the quick and sharp blows with which the bath-valet must strike the back of the *showerer.*

You must not be stingy with time and shirk the long and harsh friction that must return both heat and normal blood circulation to the body.

You must not, on the weak pretext of an unpleasant tickling, remove your foot from the boy's hands and the roughness of the linens. On the contrary: the lower extremities must be reheated quickly; you should even insist that the bath-valet who gives the rub-down also whip the soles of your feet a little. The main thing is to bring heat back to the skin.

After the friction, you should quickly help nature's work either by indulging in some energetic movements or by taking a walk of twenty minutes or a half hour; but under no circumstances should one get in the carriage to go home.

During the shower, you should not remain still. Using your hands, flog the rest of your body. Rub your arms and legs, especially your chest. The more you jump around, the better.

When you approach the overhead shower (it is usually with it that the process begins), do not place yourself under its streams gradually but all at once. The *crescendo* of seizure is an unpleasant nicety. Arm yourself with courage; the high-pressure jets will come next and console you and confirm your initial confidence.

The shower is a wonderful restorative, an unmatched modifier; but, as with all good things, it should be used only with discernment and in moderation. Hydrotherapeutic treatment must vary according to a person's temperament. Before undertaking it, one would be well advised to consult one's doctor to ask for a specific prescription, as if one were going to the pharmacist. The head of the establishment, with his vast experience, will take care of everything else.

E. PAZ,
Director of the Grand Gymnasium.[107]

VELOCIPEDE RACES

Farce

FIGURE 3.20. "Here they come".

Cast:

Isabelle, flower girl

Hortensia, wheelwoman[108]

Rosette, wheelwoman

Félicie, wheelwoman

Bernard, sportsman

Hippolyte, sportsman

Anatole, sportsman

Englishman, sportsman

Male and female Velocipedists are on the turf at the Pré Catelan. They are joined by fans, gamblers, onlookers, wealthy spectators, and journalists. It is a sunny day. Music from marching bands can be heard in the distance.

ISABELLE, *wandering through the crowd.*

Gentlemen, come and get a rose boutonniere or a bouquet of lilacs. (*To an Englishman.*) Decorate yourself, milord.

ENGLISHMAN

No, I dare say!

ISABELLE

I won't charge you.

ENGLISHMAN

In that case, I shall indeed take one.

ISABELLE, *attaching a flower to his lapel.*

There you are. Now, don't forget to tip the florist.

ENGLISHMAN

I say! I should rather return the whole kit and caboodle.

ISABELLE, *walking away.*

Oh, the English! They killed my emperor![109]

BERNARD, *to his friends.*

Indeed, my good men, it's like you say. The track was two thousand meters long with no climbs. Crafty came in first at four minutes fifty seconds riding a Velocipede with a ninety-centimeter front wheel.

ANATOLE

Amazing! What about little Thingamabob?

BERNARD

Thingamabob put his head down at the start. But try as he might, he came in second place with a time of five minutes. He made a nice race of it!

HIPPOLYTE

And Thingamajig?

BERNARD

Thingamajig took the lead early on but lost steam after the first kilometer. He cracked. Thingamabob beat him by three lengths.

HIPPOLYTE

That's unfortunate. He's well built.

BERNARD

Indeed, he's strong. Then they organized an endurance race.[110] Crafty finished the course in nine minutes.

ANATOLE

All four thousand kilometers?

BERNARD

With ease. It's too bad you missed it. There was a new model there made out of aluminum bronze. It's splendid: a Voltairian Velocipede. It looked so delicious I could have eaten it.

ISABELLE, *entering.*

A rose, gentlemen?

BERNARD

As usual, my beautiful child. How much for a handful?[111]

ISABELLE

You know I'm from Nanterre.

BERNARD

That's right. I'll give you fifty cents. Virtue is its own reward.

ANATOLE

What about the slow-riding competition?

BERNARD

It was entertaining. Ten riders set out to go three hundred meters, and not a single one made it! Thingamajig didn't know which way to turn and fell on his side ten feet from the finish. He claims he hasn't spent enough time working on his balance. Crafty was right on his heels. When he saw Thingamajig go down, he wanted to get to the line as soon as possible since he was the last one on course. But, like Telemachus, he had to go around his adversary. He turned too quickly and fell.

ANATOLE

Were a lot of people there?

BERNARD

Yes, a lot. A very elite audience. I saw Mourcheron, Toupette, and the Duke d'Olivarès.

HIPPOLYTE

And women?

BERNARD

Hordes of them; all very distinguished: Blanche, Cora, Mucous, Monkey Wrench. Monkey Wrench was wearing the Golden Fleece as a bracelet.

HIPPOLYTE

Who gave it to her?

BERNARD

A sheep, of course!

ANATOLE

Gentlemen, while you're making jokes, we're losing our place. Look at the crowd. And the wheelwomen are on their way.

HIPPOLYTE

You're right. Are you coming, Bernard?

BERNARD

Into that mob?

HIPPOLYTE

If you want me to introduce you to Félicie.

BERNARD

Lead the way.

(*They make their way through the crowd and go toward the women, who are getting ready on the turf. The orchestra plays the theme from "Le Beau Dunois."*[112])

HIPPOLYTE

Hello, ladies. Allow me to introduce you to a fine gentleman, one of my friends.

ROSETTE

Some loser with a comb-over?[113]

HIPPOLYTE

On the contrary. (*To Bernard.*) Don't be intimidated by her, Bernard. Madame has a sharp tongue. (*To Rosette.*) What's more, my dear, your criticisms are misplaced. Bernard is hopelessly smitten with Félicie.

FÉLICIE, *greeting him.*

Pleased to meet you!

HIPPOLYTE

Nothing will come of it. We all know that Félicie is in the Foreign Legion. Yes, my friends, like the Rhine, she is now in the hands of the enemy.[114]

BERNARD

Allow me to express my condolences, madame.

FÉLICIE

Monsieur!!

HIPPOLYTE

Even though we trained her! But the influence of Bismarck?

BERNARD

What? You're deserting, madame, with legs like yours?

FÉLICIE

Monsieur!!!

HIPPOLYTE

The case is closed. When a woman is too beautiful, she forgets how to speak.

BERNARD, *to Félicie.*

Would you allow me to offer you my arm while waiting for the race? (*Bernard and Félicie walk away together.*)

HIPPOLYTE

Well, too bad for Prussia! Do you have your wits about you, Rosette? Are you ready to crack the whip?

ROSETTE

Not exactly. But I don't understand how anyone can love geese!

HIPPOLYTE

Why not? The goose is poultry for the wealthy; it's even better with chestnuts.

ROSETTE

And it has to be cooked. You know who I'm talking about, don't you?

HIPPOLYTE

I have my suspicions.

ROSETTE

Speaking of Félicie, I thought you also had a thing for her.

HIPPOLYTE

Perhaps. But in the distant past!

ROSETTE

It was only two weeks ago.

HIPPOLYTE

Time flies!

ROSETTE

How dare you speak so flippantly about love while you're flirting with me!

HIPPOLYTE

Me? I'm flirting with you, Rosette?

ROSETTE

Why, yes! At least it seems like you're chatting with me.

HIPPOLYTE

That's the best thing I've heard today.

ISABELLE, *entering.*

Monsieur, a bouquet for madame?

HIPPOLYTE

Where's your head? She's about to ride her Velocipede.

ISABELLE

And?

HIPPOLYTE

You can't ride a Velocipede with a bouquet.

ISABELLE

Nor with a riding crop.

HIPPOLYTE

She's right. Why do you have a riding crop, Rosette?

ROSETTE

I don't know. Just an idea I had.

ISABELLE

Then why not take the bouquet?

HIPPOLYTE

Take it, if it makes you happy. (*He hands her the bouquet.*)

ROSETTE

Fine. (*She takes the bouquet triumphantly.*) Ha! Ha! You see, you are flirting with me!

HIPPOLYTE, *he takes Rosette's arm and leads her away.*

You think?

ANATOLE, *he has been quietly speaking with the third wheelwoman for the past several minutes.*

I'm telling you, Hortensia, it's not appropriate. If you don't know this gentleman, why do you speak with him so casually?

HORTENSIA

Because he's annoying.

ANATOLE

You should have told me.

HORTENSIA

Why, so you could argue with him? Oh, god! You, my Anatole! I forbid you to speak with Rosette.[115]

ANATOLE

I haven't said a word to her.

HORTENSIA

Yes, you did. Just a minute ago, you said, "Hello, ladies."

HORTENSIA

So?

HORTENSIA

So . . . Rosette heard you say it.

ANATOLE

You're a Hyrcanian tigress.[116]

HORTENSIA

If you speak with her, I'll make a scene.

ANATOLE, *attempting to change the subject, he looks Hortensia over from head to toe.*

Ah! I see you've been saving money!

HORTENSIA

Me? Never. What do you mean?

ANATOLE

Your skirt is too short; I can see your knees.

HORTENSIA

Well, I'm not knock-kneed, am I?

ANATOLE

That's not what I mean. You aren't hunchbacked, either. It would be great if we only covered our flaws.

HORTENSIA

My god, you're stupid! You know I'm not hiding anything from you.

ANATOLE

But you don't have to hide anything from me. That's not a good excuse.

HORTENSIA

Well, it's my excuse. Do me a favor and stop annoying me. I won't be able to race. Listen, my little cat, what will you give me if I win?

ANATOLE

Anything you want.

HORTENSIA

And if I lose?

ANATOLE

I have to give you something if you lose?

HORTENSIA

Oh, much more!

ANATOLE

Then do try to win.

(*A bell rings. The wheelwomen take their places at the starting line. The Velocipedes twitch. The women stand confidently next to their mounts, hands on the handlebars, pride in their eyes. The bell rings again, and they rush out on the track.*)

BERNARD

I've got twenty-five louis on Félicie![117]

HIPPOLYTE

I'll take that bet as a courtesy. I'll wager on Rosette, of course. And you, Anatole?

ANATOLE

I'm not wagering a thing. I'm still trying to understand Hortensia's logic.

BERNARD

Look, gentlemen! Here they come.

HIPPOLYTE

Rosette is in the lead. What a little demon!

BERNARD

You know, Félicie is really quite charming. Look at her flexibility, her elegance.

ANATOLE

Bah! She's way behind. Hortensia is coming back. She's reeling in Rosette. Bravo! We win!

HIPPOLYTE

Poor Rosette!

BERNARD

Gentlemen, Félicie is losing her head. She's not riding in a straight line. Her Velocipede is weaving all over the place. I've lost!

(The band plays. The women, out of breath, step off their Velocipedes and are surrounded by fans. They seek refuge in a clubhouse where only a few others are allowed to enter.)

ANATOLE

Hortensia, come here! Let me hug you! Fifteen hundred meters in two hundred seconds! You're an angel! Show me your wings!

HORTENSIA

Are you happy with me?

ANATOLE

Should I carry you triumphantly in front of the crowd?

HORTENSIA

You see, my skirt wasn't too short!

ANATOLE

Maybe you're right. They're calling for you. Go wave to the adoring crowd.

(*Hortensia goes to the window. Cheers can be heard, and the band begins playing the "Beau Dunois" again. She blows kisses to the public.*)

ROSETTE

Monsieur Hippolyte.

HIPPOLYTE

What would you like, my sweetheart?

ROSETTE

My hat, my jacket, and your arm. I'm irritated.

HIPPOLYTE, *handing Rosette her hat and jacket.*

Here you go. But you know I bet twenty-five louis on you?

ROSETTE

Next time just give them to me instead. Let's go, please.

BERNARD, *to Félicie.*

Oh, madame, it would be difficult to lose with more charm than you showed. Don't be upset; grace is worth at least as much as speed. You're too beautiful to take a job riding as if you were on a horse.

FÉLICIE

Monsieur!

BERNARD

Those other women have livelier legs, but yours are rounder. You can't ask an apple tree to produce roses.

FÉLICIE

Monsieur!!

BERNARD

If you trust me, let's go to dinner. Nothing is as important as celebrating defeat with a drink.

ISABELLE, *entering.*

Monsieur, here's my last bouquet.

BERNARD

I'll take it, my child.

ISABELLE

No, keep your money; it's a gift. I'm done selling flowers. These Velocipedes have stirred my soul. I'm going to start training!

BERNARD

What has gotten into you?

ISABELLE, *inspired.*

A vocation.

G. R.

FIGURE 3.21. Velocipedists converse.

THE VELOCIPEDE ASSOCIATION

Readers might find it useful to read excerpts from the statutes of the Velocipede Association, which was just founded in Paris.[118]

EXCERPTS from the legal decree approved by the Chief of Police, 22 December 1868:

ARTICLE 1

The Velocipede Association is legally authorized.

ARTICLE 2

The following statutes of this Association are approved as follows.

ARTICLE 3

The members of the Association must strictly comply with the conditions below, namely:

1. They must not make, without prior approval, any modification to the following statutes;
2. They may not admit any nonmember in small-group meetings or general meetings; and, in these meetings, they should not treat any matter that is not directly related to the Association's goals and objectives;
3. Once a year, they must provide a list of names of the Association's members;
4. They must comply with all the other conditions that the Association's officers may find necessary to prescribe, in particular in the interest of public security;
5. They must inform the prefecture of police, at least five days in advance, of the location, the day, and the time of meetings.

ARTICLE 4

This authorization may be withdrawn immediately for violating any of the above provisions or for violating any that are included later.

The rest of the decree contains standard administrative language.

What follows are excerpts from the Association's bylaws that may be of interest to our readers.

ARTICLE 1

The purpose of the Association is:

§ 1. To establish relationships between all those who are interested in Velocipedes.

§ 2. To study which newly invented riding systems are the best available.

§ 3. To encourage the construction of new models.

§ 4. To organize races and exhibitions and award prizes.

§ 5. To spread enthusiasm for the Velocipede and, using all the Association's means, to highlight the usefulness and the pleasure of riding.

§ 6. To buy, on behalf of Association members, Velocipedes, which it will resell to them in line with conditions determined by the board of directors.

§ 7. To lease Velocipedes purchased by members on credit.

ARTICLE 2

§ 1. The Society is composed of full members, who pay an annual subscription of twelve francs, and life members, who pay a one-time fee of 150 francs.

§ 2. Regular members and life members will have the right to enter all races and exhibitions. This right is for the individual, and it is not transferrable.

ARTICLE 3 relates to the hierarchical organization of the Association, to internal housekeeping matters, and to its administration.

ARTICLE 4

§ 1. The Association holds its meetings every two weeks, but members can meet for rides and for comparisons between the various Velocipede systems; however, no awards will be distributed unless, by a vote of the Association, such gatherings are formally recognized as Association proceedings.

§ 2. Meetings are chaired by the president, in his absence by the vice president, and in their absence by the member in attendance who has received the greatest number of votes.

§ 3. Members of the committees in charge of testing the velocipedes must write down their opinions on cards, each one separately, and the committee secretary will be responsible for reading out the cards' contents in the next session, without naming the members who wrote them.

The Velocipede Association is
directed by Mr. Émile Royer, 40,
rue de Buci, Paris.

A VELOCIPEDE'S LOVERS

FIGURE 3.22. Danger at night.

I

Clémence was pudgy and brunette.

The Velocipede was thin and yellow.

She would trot down the streets, with small sways full of grace and fancy.

He would gallop down the avenues at a good clip, full of vigor and gracefulness.

Sometimes, their paths would cross at a bend in the road.

Clémence would blush—without knowing why.

The Velocipede would salute—without knowing how.

With red cheeks and salutes, both of them began to make a habit of these meetings: at first happenstance, then sought after, and then the habit turned into a need.

And therefore, whenever Clémence would not see the Velocipede, and whenever the Velocipede would not see Clémence, she felt sadness in her bones and sighs in her throat for the rest of the day; he had a whole day's worth of melancholy and "not knowing what to do."

And yet, who among us does not know the dangers of "not knowing what to do," the absurd inspirations it provokes, the nonsense it creates, and the embarrassment it causes?

Not knowing what to do transforms a chaste man with an even temperament into an intrepid Lovelace, spontaneously lavishing ardor and enthusiasm on the first bun of pulled-back hair that walks by, without the slightest bit of encouragement, without the shadow of a pretext, without a hint of sincerity.[119]

II

Their relationship continued in this way, until an entire week passed without our rider being able to lay eyes on his accustomed wanderer.

Imagine his dismay!

And yet, despite the anxiety and gloom that these successive disappointments could have created, he had—for a whole week, and thanks to the whims of happy coincidence that always distracted him at just the right time—he had, as we were saying, managed to avoid any piqued elation in the area of his member, where he experienced sweet sensations in the presence of Clémence.

Seven days! It was, frankly, a completely unexpected result; consequently, when he thought of the amount of resistance that he had shown on this occasion, our dear Velocipede felt storms of pride well up in him, greater even than those felt by Archimedes when he exclaimed, running naked on the pavement of Syracuse's boulevards, the great *Eureka!* of his fulfilled dreams.[120]

Alas, unfortunately the seven days were followed by an eighth. It always happens that way!

III

Moreover, here are the facts:

Bleak and with brakes applied, the two-wheeler was rolling down the avenue des Champs-Elysées, without paying too much attention

FIGURE 3.23. "Head over heels".

to where he put his wheels, when a simple pebble, thrown by chance, fell in his path.[121]

Nineteen times out of twenty, at any other time of day, the Velocipede would have gone over the obstacle, without even so much as noticing its presence, but that day!

In short, the poor rider—"made of metal and wood"—went head over heels, landing flat on his back.

A cry rang out.

A woman rushed to help.

When our hero opened his eyes again, he saw a head full of blond hair bent over his face, in the middle of a huge circle of curious onlookers.

For twenty minutes, in fact, the aforementioned hair, crouching near the Velocipede, attended to it with the most delicate and intelligent care. With smelling salts on her fingers and waves of sweet words on her lips, she managed to revive the senses of the poor fallen one.

Now, who was this woman?

I'll bet a hundred to one that at this question, a choir of readers will triumphantly exclaim:

"Clémence!"

Actually no, it wasn't Clémence!!!
Clémence always wears a pink hat.
The kneeling woman had a purple hat.
Her name . . .
Well, let's not get there too quickly.

IV

Once the fainting completely passed and the salts were used up, it was up to the recently revived to produce expressions of gratitude whose warmth and energy I'll leave to your imagination.

The downside is that in such circumstances it is nearly impossible to give thanks without looking, if only a little bit, at the person you are talking to.

The two-wheeler therefore looked: the purple hat happened to be adorable.

An ideally slim waist, effortlessly tightening to forty centimeters, below a pale, long face, itself topped by a certain number of kilos of Bismarck-brown hair; eyes cut into the hull of a ship, with artfully blackened eyelashes and gently detailed by hand; a little nose raising its pink nostrils with a very cute cheekiness; a mouthful of slightly thick lips but of so fine a flesh in its sparkling redness that even tiny pomegranates were not more appealing to a hungry traveling pilgrim.[122]

What else shall I tell you?

The throng of onlookers, crowded around our two characters, became quite bothersome as the continued outpouring of emotion had to be tempered because of the audience that they didn't want there in the first place.

To make a long story short, the Velocipede took the purple hat by the arm, and soon afterward they were both seated in a private booth in the Café Anglais.[123]

Our readers certainly could not expect us to listen in to the pair of diners during their gastronomic tête-à-tête.

We will just dare to be intrusive enough to:

1. Glance at the menu and simply note these four lines:

Bisque soup.
Bordeaux crayfish.
Partridge with truffles.
Chilled Ruinart champagne.

2. And listen just for a moment, a second, and pick up this snippet of dialogue that we offer as a prelude of what is to come:

"Your name, so that I can bless it in my dreams?"

"Cora," she replied.

V

During these events, what had become of Clémence? Why had she so often missed the unspoken date with her two-wheeled friend?

The answer is very simple, alas! And just as prosaic.

She had, the last time they saw each other, in the season's cool temperatures, caught a magnificently voluptuous head cold. Her nose, her cute-as-a-button nose, had gradually taken on the hues and contours of a tomato.

The poor child tried all methods known and unknown to get rid of this inconvenience as quickly as possible, an inconvenience that she worried about and that an unkind, less interested—or less polite—person would have found absolutely ridiculous.

But the more she breathed in gray salts from the palm of her hand, the more she dabbed the inflamed parts with cold cream and camphor, the more she cushioned it—if I may put it that way—with marshmallow-water fumigations, the more she spent on remedies prescribed in the pharmacopoeia, or remedies from "old wives' tales"—the less she actually stopped the subversive flag from being raised on her face.[124]

We will not spend time describing the tears, impatience, despair, and rage that was spread, manifested, felt, and foamed, in the midst of vials, pots, and boxes of all sorts that she unnecessarily piled up around herself.

Her most focused thoughts were forever stretched and pointing toward the moment at which, every day, at the same time, the rims of the Velocipede were spinning, yellow and dizzying!

Sometimes she tried to sew, sometimes she read her fortune in tarot cards, sometimes she wanted to read, other times to tinkle at the piano.

And yet, on her embroidery of intricate arabesques, on Mademoiselle Lenormand's illuminated figures, on the black regimented lines marching through the white snow of her novel's pages, and on the yellowed ivories of her Érard, her sneezes fell in a fine and repeated mist, without any signs of letting up.[125]

On the evening of the seventh day, the annoyed patient broke all her boxes, smashed all her vials, threw out all her water, burned all her herbs, and, suddenly, gave up all medication.

The next day, the noisy and weeping head cold had moved away—and with it, all the tears and noise—and the nostrils returned to their ordinary shade; unfortunately, by the time she noticed it—you can imagine with what joy!—it was too late.

The usual meeting time had passed.

She would have to wait another one thousand four hundred and forty minutes!

VI

"No matter," she said to herself; "I will go out all the same, if only to distract myself from the cruelties of waiting."

Clémence—raised according to English customs—went out alone; moreover, as she was an orphan, and therefore independent, she had always had the good mind not to give up her freedom of movement: not for anyone. With good judgment and a sound mind, she had, until then, gone through life with sure footing, resolved to stumble only on a stone of her choosing. When, on occasion, people tried to put themselves in her path, she turned away, shrugging her shoulders, scowling, laughing, or smiling with disdain, as the case called for.

However, her twenty-four years were beginning to nag her with their demands for intimacy when she saw our Velocipede for the first time.

And, back in her room, she had dreamed for hours, and her chest had swelled with previously unknown sighs, and—when it was time for bed—she could not close her eyes and all night felt as if her mattress was stuffed with rose leaves.

If it had been so the first time they saw each other, we can easily imagine what it must have been like after the hundredth!

These things, in fact, had already been going on for three months when the head cold threw itself across her path.

And—unbelievably!—during these three months, nothing was shared between them other than what we have already mentioned, namely:

Sunburns and polite greetings.

VII

So she put some effort into her appearance—a lot of effort, in fact—and went out.

Was it intuition that directed her path? *Chi lo sa*?[126]

But what is certain is that at one point, she suddenly froze—like a pointer—at the door to the Café Anglais.

It was him!

~

He is on the sidewalk, back turned, hat off, face animated, and mouth full, chatting with one of his friends, a gentleman.

Clémence's heart is beating enough to burst the seams of her corset. Trembling, her eyes burning, she is stuck there on the sidewalk, rigid and pale as marble, staring at him.

This strange and prolonged posture ends up drawing the attention of the aforementioned gentleman, who expresses his astonishment; the intrigued Velocipede turns around . . .

If ever the literary convention of replacing description or psychological analysis with a short outburst followed by punctuation signaling shock—*"Scene!"*—could be justified, it is most assuredly in this circumstance.

VIII

At first speechless, the two-wheeler recovers after a moment, rushes toward Clémence, then changes his mind, returns to his very stunned friend, scribbles three lines on a sheet torn from his notebook, folds it, attaches a hundred franc note to it, and puts it in the gentleman's hand. At the same time, he takes his friend's hat, places it on his own head, says a few words in his friend's ear, then pushes him toward the restaurant's staircase. Finally, with a smile on his lips, his arm curved, he approaches the young woman, and . . .

And, while the gentleman slowly climbs the steps of the fashionable restaurant, Clémence and her Velocipede, leaning on each other and chatting in low voices, walk away quickly.

Very quickly indeed, because, a split second later, one of the windows of the Café Anglais was flung open and in it appeared a purple hat that scoured the whole boulevard with a wrathful eye. Meanwhile, from inside, a man's voice could be heard:

"But I swear to you, my beloved, that he has just been surprised, as he says in this note, and kidnapped by the countess, his wife, from whom he could not free himself."

The vexed mauve hat was obliged to answer:

"It is true that I do not see him! My word! Never mind; let's finish dinner, my dear."

And the window closed.

IX

Two months later, in a lovely boudoir in the rue Maubeuge, a dashing courtesan unfolded the *Petit Journal* and, shaking, suddenly said,

"Ha, that's a good one!"

"Which one?" asked a gentleman casually stretched out on a couch and completely absorbed in solving this serious problem:

Given the toe of a boot and the end of a walking cane, with small strokes, mold the contour of one into the plane of the other.[127]

"You ask me which one? Listen to this!"

And the beauty began to read: "Yesterday, everyone at the église de la Trinité celebrated the marriage of Count Alfred d'Orfraie with Mademoiselle Clémence Paraguay.[128] All of Parisian high society knows that the count is the most skillful Velocipedist in Europe. It is even said that he owes his happiness and his fortune to this ability . . . his happiness, because his wife is charming; his fortune, because she adds to her husband's golden means the cool million that she has just inherited.[129] We can also affirm that one of the most precious jewels he gave his new bride is a special golden Velocipede in miniature, with springs, brakes, etc., made of brushed silver, and the hubs, grips, etc., made of diamonds."

"So what?" said the young man.

"So what!?!" she resumed. "That count is the one who gave me the slip at the Café Anglais."

"And?"

"And . . . he wasn't married, and you, who told me, you stopped me from going after him!"

"Egad!"

She pretended to be angry; he got up gently, kissed her on the nape of the neck and, in his sweetest voice, asked:

"Are you really mad at me, Cora?"

She paused to think for a moment; but suddenly, and seemingly from an emotional outburst that was impossible to contain:

"Well, no, to the contrary!"

Her mouth stopped there, but her mind raced on:

"But . . . it will cost you dearly!"

X

Clémence and her Velocipede lived happily from then on.

It is true that their household is still only half a quarter moon old.

Will they have many children?[130]

Since this is a private matter, we dare not answer.

Jules DEMENTHE[131]

FIGURE 3.24. Happily ever after?

FASHION AND VELOCIPEDES

FIGURE 3.25. Fashion and velocipedes.

Brummell, it is said, changed clothes up to five times a day; and now inept dandies, burdened with debts and who have little to do, follow his example.[132] His fastidiousness in this regard may have been a bit excessive, but he argued convincingly that the clothes a man puts on when he gets up in the morning cannot be the same as what he wears when he eats, when he goes out, when he has supper, and when he goes dancing. We will not argue the point.

What remains to be determined is the outfit for the man or woman who rides a Velocipede. It is no small matter, and we recently put the question to several tailors, who debated it extensively. Based on their experience and quick observations of our own, we present the results of our thoughts on the matter to our readers.

IN THE MEN'S DEPARTMENT

No stove pipe hat; it will distract and get in the way. It can be knocked over by a tree branch or fall off because of a bump in the road; in fast races, the air resistance can be enough to whip it off a rider's head. We must therefore be satisfied with either an elegant cap or a round hat, secured with a strap.

Simple, tight-fitting clothing, with room in the thigh; pants tucked into standard or low-cut boots, giving the foot full range of motion: these are our prescriptions. We do not recommend an overshirt, unless it is very short. The seat of the pants must be solid and resistant.

IN THE WOMEN'S DEPARTMENT

Here, we are a bit torn, and we hesitate between draconian recommendations and the indulgence that is necessary for the fairer sex. Therefore, we will be less absolute in our advice; readers can choose as they please from among the following motifs:

THE KID. Short blouse, cap with a low visor, belt; pants tucked into low-cut boots. This is certainly the most convenient outfit that ladies can adopt. However, there is nothing very graceful about it, and women look very mischievous when they ride around dressed this way. It goes without saying that they cannot wear petticoats or anything else that would make their clothes larger than necessary.

THE DANDY. This costume does not need to be described. Let's just say it is absurd. We are very sorry for the thousands of Parisian women who think that the only fun thing about Carnival is putting on men's clothes. It is quite simply atrocious and in bad taste.

THE FANTASY. This gallant kit is certainly the winner; its only drawback is that it is indescribable. For the head and the torso, a Russian toque with a feather sticking out and a one-piece bodysuit with piping or lace on the edges; but the rest . . . Ah!!! The rest is difficult to describe, and it would be best to refer to our drawings. Some will perhaps find them a little too short; these prudes should take a step back and relax. There's nothing wrong with wearing lace tights that go to the knee and expose the bare leg below—but the leg must be visible.

Women's races are probably organized for the particular attraction that they represent. If women dressed as hoodlums, it would undermine the purpose of the race. Such races must display a measure of grace and elegance that depends above all on the women's clothes, the riders' flexibility, and their composure on the saddle. They probably need boots, but not high equestrian boots. The skirts must not descend too far; they should be free, floating, unstarched, and without any kind of crinoline. There is no harm in shortening them or wearing a page's uniform. It all depends on dispositions . . . and figures.

Let's go no further, because we do not want to quarrel with either Mr. Dupanloup or Mr. Veuillot.[133] Is it really so wrong to show one's legs? I have known very respectable women who say it's not. It is true that they all had very nice legs.

A TAILOR
Who will soon make himself known.

WHERE A VELOCIPEDE LEADS

FIGURE 3.26. "Are you coming with us?"

Here is the disturbing story of Count Raoul de Rochefort—who lived as a bachelor until the age of thirty and ended up with the Velocipede—as told at the Jockey Club between two hands of *écarté* while sitting around a fire and puffing on cigars.[134]

Raoul, vigorous and sturdy, healthy in body and mind, enjoyed a fortune that enabled him to live the high life. Gifted with exceptional intelligence, he realized that pretty boys and their vapid existence were no match for his temperament, no matter how pretty they were.[135] He didn't care much for the queen of hearts and wasn't giving money to any flower girls. After some hesitation, he decided to travel, to see if the world stretched beyond the place Pigalle and the quartier du Luxembourg, a notion he had been taught in school and that he had read in a few treatises of geography. He had not been lied to. He explored America, Australia, and the Grand-Montrouge suburb of

Paris, where he saw some truly curious things that he could not have encountered in the heart of Paris, even if he had paid to see them. After a few years of adventure, he volunteered for Dr. Pettermann's expedition to the North Pole, less to arrive at the top of the earth than to discover for himself the linguistic faculties that we may have a little too liberally attributed to seals. He was admirably and very prudently prepared and equipped, but nostalgia got the better of him when he was about to buy furs on the boulevard Montmartre. He yielded his right of passage to Jules Verne, who wanted to verify some harsh terrain from the manuscript that Captain Hatteras had bequeathed to him.[136] Deciding to become Parisian again, Raoul thought of paying a visit to his uncle.

Count Raoul's uncle had the title of Marquis Gideon de Marcheprime. Sixty years old, with a keen and clear eye, a quick mind, and easy mannerisms, the old gentleman could have lied about his age, were it not for the fits of gout that kept him chained to his chair. He received his nephew without difficulty, although Raoul was his heir, and he saw in his nephew the defects and qualities that had once made himself successful in the world.

Raoul—and this may seem extraordinary for the hero of a novel—did not lack flaws. He was an ardent spirit, but he was enthusiastic only when he had good reason, and he did not take peanuts for diamonds no matter how good they might look.[137] He shone in the highest levels of Parisian society and made the best of Arsène Houssaye's hiding places, where, with his good looks, he had seduced the queen of a neighboring country, recently dispossessed.[138] Discretion prevents us from saying more. Raoul behaved rather badly under the circumstances, for he had traveled so much and seen so many black, white, yellow, and red women that he knew a thing or two, and should have known better. He typically weighed women on the scales of youth and beauty, which led him to making mistakes a little less often than others.

He never bothered with men. He found them so perfectly ugly that he could only barely make an exception for himself. For this reason, he had never quite understood why women decided, from time to time, to love the virile monkeys who represent half the human species. But he had to face the facts: in women's defense, it may be said that man is, on the whole, the cleanest of orangutans, and, in this respect, women make the best out of a bad situation.

Heading toward the rue du Bac, where his uncle the marquis lived, Raoul was not free from worry. He had not seen him for a good ten years and had sent only two or three letters reminding this kind old man that somewhere in the world he had a nephew who was counting on his inheritance. Nevertheless, Gideon recognized the young man at first glance, and he held out his hand firmly with no other display of tenderness, for emotions were foreign to him. The count was about to begin the tale of his adventures when the old man interrupted him, to tell him, for the one hundred seventy-ninth time, about the taking of the Trocadero, where he had had the distinguished honor of commanding a company of infantrymen.[139]

This terrified Raoul, who outwardly remained composed and who remembered to cheer at the right parts of his uncle's story. Seeing this, the old man took him into his confidences and admitted that he had recently disinherited him in favor of a twenty-year-old girl who claimed to be more or less his niece and who had become, for better or for worse, his guardian angel.

"But of course you know her," he said. "It's little Pipette."

"Pipette!" answered Raoul, sitting up with surprise. "That red-faced monkey I saw rolling around in your hallway ten years ago? That's no woman! And at any rate, my dear uncle, think a little: Wasn't that the daughter of Coraline, my late aunt's chambermaid?"

"Indeed."

"So then how could she be your niece?"

"Ah!" said the uncle, a little embarrassed. "There are things like this that are surprising at first sight. What is certain is that she is a charming girl. She dotes on me, she amuses me; she says that my rheumatism suits me well!"

"You're being taken advantage of by a little minx!"

"Huh?" said the uncle, frowning. "You don't know what you're talking about, dear nephew. I've got no problem with you running around in Australia. But for you to tell me that I'm wrong to prefer, to you whom I never see, a flattering, pretty, and obliging young girl—well, it pains me."

"It hurts me, too, Uncle. Be careful; the two of you won't be together long."

"You're a scoundrel. Pipette is virtuous, and I'm less perverted than you think. I don't really want to disinherit you! Pipette is very respectable."

"What do I care?"

"She is charming and distinguished; she hits a shuttlecock like dearly departed Racquet himself. Why not marry her?"

"Pipette?!?"

"Yes, Pipette."

"Hell's bells!" At this point Raoul was about to swear like a sailor. But, upon reflection, he stopped. A smile curled the corner of his lips; he slowly smoothed the ends of his mustache.

"That's an interesting idea you've got there," he said.

"Before you get upset, think about it for a moment. I have reason to believe," said the old gentleman, smugly rocking back and forth, "that Pipette is closer to me than you might think. She persuaded me of that, in fact. She cares for me; she's fattening me up so much that I'm beginning to feel like she's already part of the family. So, if she's my daughter, you wouldn't be marrying down."

"To be honest," said Raoul, "that's the least of my worries. Beautiful hair, big eyes, and little white teeth will always be the purest sign of feminine aristocracy. No titles of nobility are worth the brilliance of two beautiful shoulders, covered in lace and offering views of peaks and valleys. And as for a coat of arms—"[140]

"You shouldn't talk about it," said the uncle, "unless you want to see my gout act up again."

"I'll drop it, but I could go on and on!" said Raoul. "You can see that I'm keeping an open mind. But let me just say about Pipette: may the Devil take her! . . . Well! No, actually, he won't take her! I'm telling you: Pipette is as black as a mole, ugly as an orangutan, and a little hunchbacked, too."

"I get it!" said the uncle. "Do you really think I have such bad taste or that I'm blind? It's one thing to see a ten-year-old girl, scrawny and sickly, and quite another to see her at twenty, fresh and flowery, well put together—and, if she has humps, they're not in her back. Besides, it won't hurt to have a look. And actually, it's time for our trip to the woods."

"You're going to the woods?"

"Without a doubt."

"And how?"

"In a wheelchair."

"And Pipette?"

"On a Velocipede. Are you coming with us?"

"No way!"

"As you wish."

At that moment, Pipette entered, dressed in a light, elegant suit that recalled the uniform of the bloomers-wearers, corrected by feminine coquetry and French taste.[141] High-heeled boots molded to her arched feet, and thin legs disappeared at the start of her calves, which got lost in a sea of lace. Pants or petticoats, cascades of fine linen and embroidery undulated under a short skirt, which hardly went much lower than the knee. The top of the garment was inspired by the waistcoat and the riding dress, but instead of the stiffness of flat collars and starched shirts, the young woman's neck, of a golden-white tone, was covered with a tie that was very loose and wide, made of black satin with cherry accents. Put a little round hat on top of a quick, mischievous, somewhat insolent face lit up by two large black eyes and you have an idea of the apparition that rendered Raoul mute for a few seconds.

"The weather is clear," said Pipette, "and the sun is shining. The woods must be glorious. Are you coming, my old friend?"

"Without a doubt," said the amiable octogenarian, "but I would like to convince my nephew to join us."

Pipette finally seemed to notice the presence of a stranger. She scarcely turned and bowed to Raoul, who had bowed.

"Mr. Raoul?" she said. "I remember you very well. Do join us."

"I would certainly come, miss, if I had a horse to ride."

"I can offer you a Velocipede, if you know how to use it."

"I admit that I have never tried, but I have some idea of the calisthenics involved."

"Oh! Oh!" said the girl.

> Men like you need not attempt a new pursuit twice,
> And achieve masterstrokes without any advice.[142]

"Understood; I will go get our mounts readied."

She went out haughtily, the cane in her hand whistling.

"Well!" said Uncle Gideon. "Isn't she a wonder?"

"Ha, ha," said Raoul, quite restrained. "There's a lot I could say about her. You see, Uncle, I'm suspicious of your Parisian beauties. They're slight, fragile, and wispy; you can't get your hopes up much for that. They're like puffed omelets. When they're dancing and moving around, they fan their tails out like peacocks in the sun; their eyes shine, their voices ring out, they are really very pleasant. And then, at the first cool breeze, what a letdown!"

"Do you think Pipette is like that?"

"I don't know what to believe. She is less ugly than I thought she'd be. But I'm an experienced traveler. Look, I met a queen in Australia who, with only one necklace to her name, still found a way to convince people she was royalty."

"You surprise me!"

"I even got caught up in it myself. No, I don't let myself fall for the appearance, the allures, and the smiles of your dolls.[143] I'm not that stupid!"

"At any rate," said his uncle, "everyone's entitled to their opinion. Let's go!"

FIGURE 3.27. Pipette.

Half an hour later, Pipette entered the Bois de Boulogne, riding circles around Uncle Gideon's chair as he basked in the sun. After a short lesson, Raoul was following her closely, showing only a few signs of a beginner's inexperience, just like back when Elleviou, needing to tear himself away from private life and throw himself onto the stage, became in very short order the leading actor of his day.[144]

Raoul, however, was distracted, and his progress clearly suffered for it. Pipette, straddling the Velocipede, caught the attention of Raoul and everyone else. Raoul watched her with interest and began to wonder if it wouldn't be so bad after all if he gave in to his uncle's whim, even if it meant admitting that Pipette was not hunchbacked. Based on what the marquis had said, he took her to be virtuous and well educated. There were some points in this story that worried him and others that intrigued him. It is not that he cared so much about the inheritance, but there are certain mindsets that the most foreign ideas infiltrate ever so slowly, and the idea of marrying, which had at first so strongly repulsed him, no longer struck him as so strange. All that remained was the choice of the woman: Pipette was certainly as good as another, or at least she seemed to be.

How could he be sure? Driven by overzealous gallantry, Raoul committed one reckless act after another: he fluttered around the young girl, admiring her slender, graceful bodice that swelled as she breathed and her little feet's vigorous movements. He was surprised to see legs so round for such a slim waist.

Amid this daydreaming he completed a vast loop calculating neither the diameter nor the radius. Instead of rolling up alongside the young woman, he was heading straight for her, and so he let out a shout; she saw him and tried to get out of the way. Too late! Both going at full speed, they met at right angles to one another and went tumbling in the dust.

The hardest impact was fortunately borne by the Velocipedes, which were smashed to pieces. Raoul got up in the blink of an eye, but Pipette, curled up like a shot hare, was thrown to the turf in an awkward heap.[145]

Raoul took her in his arms and, jumping into a passing carriage, had them taken to his uncle's residence at full speed. His uncle remained alone, in the midst of the debris of the Velocipedes.

Uncle Gideon went home, mildly anxious, after having given his name to the guards at the Bois de Boulogne, who said they had to file a report of what had happened. His first thought was to go upstairs to his niece's room, where he found her lying on a chaise longue, a little pale and with her clothes extensively unbuttoned.

Raoul was with her; he greeted his uncle with an outpouring of warmth:

"I'm marrying her!" he said.

"But," said the uncle, with a little spite, "what about the titles of nobility?"

"She's plenty noble for me."

"And her coat of arms?"

"I'm happy with it."

Sometimes, everything just *falls* into place.

So ends the legend of Count Raoul de Rochefort, who lived as a bachelor until the age of thirty and ended up with the Velocipede.

A. R.

ON CHOOSING A VELOCIPEDE

FIGURE 3.28. Two velocipedists.

We have no specific advice to give our readers on the choice of a Velocipede, other than to contact a reputable manufacturer, such as the one that we recommend later. A poorly assembled or badly made Velocipede can indeed cause serious accidents.

Velocipedes must be constructed of top-quality cast iron, and they must meet special conditions for being both strong and lightweight. Their weight varies from twenty to thirty kilograms, without panniers or lights. The ratio between the diameters of the front and rear wheels is usually 80/65 cm, 85/70 cm, 90/75 cm, 95/80 cm, or 100/80 cm.

Wheels that are larger than a meter in diameter are used only as the large wheels of three-wheeled Velocipedes, which are actually cabriolets.

When ordering or buying a Velocipede by mail, it is useful to send the manufacturer the rider's overall weight and leg length so he can build a machine with the most comfortable dimensions and the necessary sturdiness.

N.

PODOSCAPHS, OR MARITIME VELOCIPEDES

This little book would be incomplete if we didn't devote a few lines to Podoscaphs, some models of which floated on the river in the Bois de Boulogne in 1868.[146] In the middle of a long, narrow boat sits a wheel that is about a third under water through an opening in the bottom of the boat. The opening has flaps around it so that water won't enter the boat. Straddling the bump that covers the contraption, the navigator puts his feet on pedals attached to the wheel hub and applies a rotational movement. The wheel is equipped with paddles that beat the water and quickly move the boat forward.

It seems that this mechanism could be simplified by replacing the one central wheel with two wheels spread apart like on old steamboats. They would be united by an axle to which single or double pedals could be affixed so that two people could apply power to the motor together.

It would also be easy to use human force to turn a propeller placed at the rear, pushing the boat forward. It would simply require coordinating the transfer of movement from either the center of the boat or from wherever the force would be most easily generated.

Such modes of locomotion are related to Velocipedes in that they require effort analogous to that which takes place in the act of walking. This force is not derived solely from the muscles; part of their movement also comes from the weight of the body, or at least of the lower limbs. It is therefore advisable to place the Podoscaph pedaler in a slightly elevated position; while it makes the boat less stable, one would get more out of each pedal stroke. Thus, in this activity, there are precautions to take and dangers to avoid.

P.

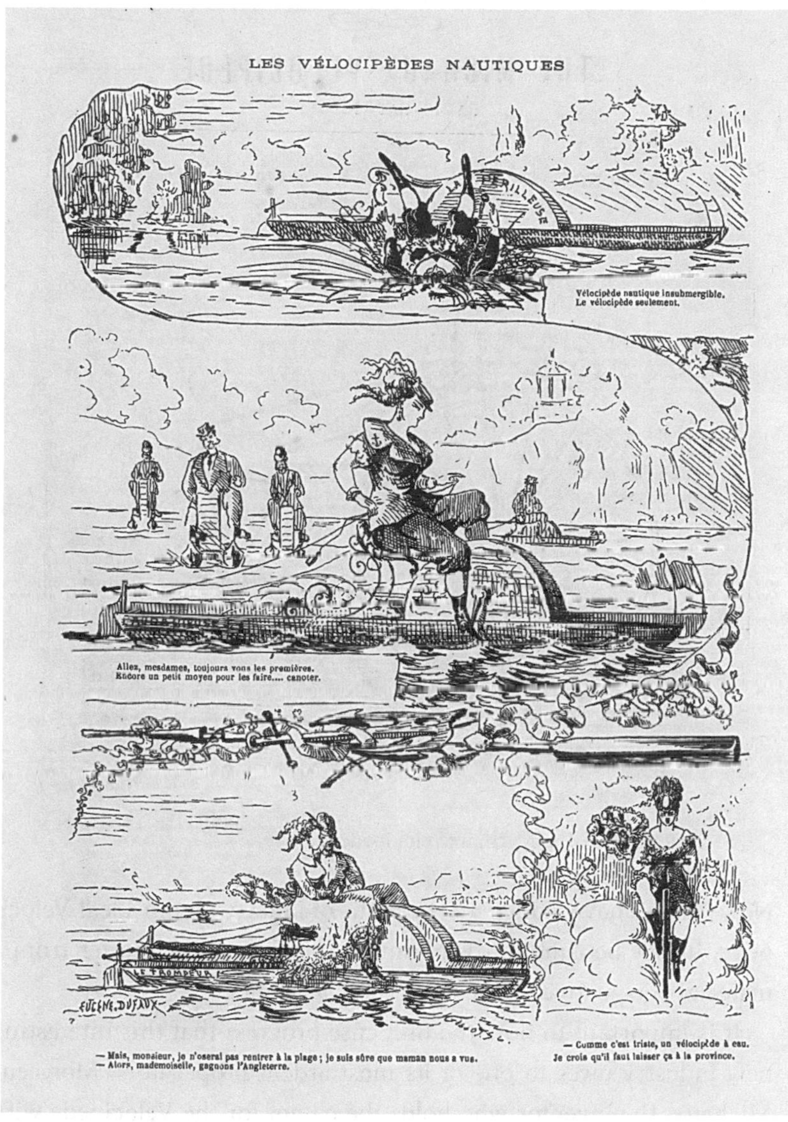

FIGURE 3.29. Eugène Dufaux, "Nautical Velocipedes," *Paris-Caprice*, April 17, 1869.

THE MICHAUX VELOCIPEDE

FIGURE 3.30. The Michaux velocipede.

Now that we have ridden all the paths of fantasy on our ideal Velocipede, it may be time to dismount and deal with the matter from a material perspective. Prose never loses its rights.[147]

It is important to note the immense progress that this interesting new industry owes to one of its most ardent propagators, Monsieur Michaux, the inventor who holds the patent for the Velocipede with pedals, and whose name is synonymous with the best manufacturing of the most deluxe, useful Velocipedes.

The Michaux factory and workshops, located at 27, rue Jean-Goujon, in the Champs-Elysées area of Paris, employ a considerable number of workers.[148] The whole industrial site, whose success is increasing day by day, also includes an elegant school where Velocipede enthusiasts can learn, in just a few lessons from a skillful teacher, the art of maneuvering their mount and keeping their balance.*

Members of the public are constantly visiting: mothers, friends, worried souls who hesitate before approaching the Velocipede—they all come to watch the young people's lessons and to observe their rapid progress, since uniquely skilled students can become skilled riders in just a few hours.

The Velocipede with pedals that Mr. Michaux created, and which appears at the beginning of this chapter, stands out among all others for its elegance, its light weight, and its sturdiness. This explains and justifies the tremendous vogue that it enjoys with connoisseurs. It is the thoroughbred compared to the carriage horse. Monsieur Michaux has shared with us the relatively modest prices at which his Velocipedes are sold, and we believe that this information may interest our readers:

VELOCIPEDES

Velocipede of thin iron, painted, with iron pedals, bronze hubs, brakes, patent leather saddle: 270 francs.

Velocipede of thin iron, painted, bronze hubs, crank arms hidden in the hub, padded handlebars, bronze pedals, calfskin leather saddle: 300 francs.

Velocipede with polished fittings, painted wheels, patented pedals, and self-lubricating hubs: 400 francs.[149]

* Michaux and Company, manufacturers of Velocipedes: factories, manufacturing workshops, school and testing arena, located at 27, rue Jean-Goujon, in Paris (with factories in the provinces). Not to be confused with other houses of the same name.

ACCESSORIES SOLD SEPARATELY
Genuine Michaux Components

Pedals (Michaux patent), 35 francs
Lubricating reservoirs for hubs, 10 francs
Coat rack, 10 francs
Universal wrench, 6 francs
Panniers for touring, 20 francs
Patent leather saddle, 10 francs
Padded handlebars, 20 francs
Various bags, 10–30 francs
Rubber saddle, 30 francs
Pigskin saddle, 15 francs
Oil cans, 5 and 10 francs
Lights, 10–30 francs[150]
Bronze pedals, 20 francs
Concealed cranks, 20 francs
Odometer, 30 francs[151]
Rubber grips, 20 francs

Michaux and Company accepts orders for Velocipedes designed for distance, leisure rides, obstacles and performances, and races, or children's Velocipedes at a fair price.

To place an order, indicate leg length and preferred paint color.

Three-wheeled Velocipedes have the same system as that of two-wheeled Velocipedes, differing only by the addition of a second rear wheel, priced at 400 and 450 francs.

Michaux Velocipedes and their accessories come with a full warranty. All raw materials used in the manufacturing process are carefully selected.

Three-wheeled Velocipedes can easily be transformed into two-wheeled Velocipedes by simply changing the rear axle.

We note that Michaux has representatives in major French cities and abroad.

Michaux's Velocipedes with pedals can be found on major boulevards in Paris, Marseille, Vienna, and Berlin, as well as on all the main provincial roads. His machine carries the rural postman at full speed, thereby reducing fatigue and speeding up service. With a Michaux Velocipede, journeys are no longer tiresome and protracted; trips of several leagues are transformed into pleasant strolls. It removes distances, strengthens friendships, can be stored in a corner, and never costs anything in feed.

NOLY.[152]

FIGURE 3.31. The velocipede "strengthens friendships."

CHAPTER FOUR

VELOCIPEDOMANIA IN VERSE

THÉODORE DE BANVILLE, "HERE COMES VELOCIPEDE MAN" (1868)

The first poem devoted to the velocipede, titled "L'Homme vélocipède" ("Here Comes Velocipede Man"), was written by Théodore de Banville (1823–91). One of the leaders of Parnassian poetry of the 1850s and 1860s and a master of poetic form, Banville had made a name for himself through his collections *Les Cariatides* (1842), *Les Stalactites* (1846), and *Odes funambulesques* (1857), as well as his contributions to the 1866 volume of *Le Parnasse contemporain*, which anthologized the leading poets of the day. In "Here Comes Velocipede Man" (1868), Banville chose the fixed form of the triolet (also called the *rondel simple*): its cyclical nature is fitting for a poem about the velocipede. Traditionally, a triolet's first, fourth, and seventh verses rhyme with each other, as do verses two and eight. In this lighthearted octet, Banville suggests that a man riding a velocipede will make for a new creation—half human, half machine—that will defy the work of naturalists like the Comte de Buffon (1707–88; see the preface of the *Manual*) and Bernard Germain de Lacépède (1756–1825). Their exhaustive *Histoire naturelle, générale et particulière* (begun by the former, completed by the latter) had been published in thirty-six volumes from 1749 to 1804, a century before Alfred Jarry would blur the line between cyclist and machine in his novel *Le Surmâle* (1902). The tone of Banville's triolet is decidedly less ominous and its original title, "L'Homme vélocipède," also playfully hearkens back to the title of Julien Offray de La

Mettrie's controversial 1747 work, *L'Homme machine*. La Mettrie argued that, like animals, like watches, and like automatons, man, too, is a machine. Banville, however, suggests that man has willfully integrated himself with a machine, allowing him to escape Buffon's ideas of species degeneration and immutability and to evolutionarily jump ahead, becoming an "animal nouveau" (new creature).

Half wheel and half brain,
Here comes Velocipede Man.
He glides, gentler than a lamb,
Half wheel and half brain.
This new creature makes light
Of Buffon and Lacépède!
Half wheel and half brain,
Here comes Velocipede Man![1]

RAVENEL (ATTRIB.), THE VELOCIPEDE'S POLITICAL INCLINATIONS (1869–70?)

Another short untitled poem about the velocipede is attributed to the critic Ravenel and dated to this period.[2] Situating the machine in the political arena, it offers the reader a cautious reminder of the consequences of leaning too far to one side, and of the need for maintaining one's balance; forward, centrist momentum is easier to maintain for citizens and governments alike:

To master flexibility
In the grand art of politics
And shore up one's ability
A velocipede's the fix.

On a saddle that is so slight
One must quickly learn to lean
Either to the left or the right
While staying centered in between.

To maintain perfect balance,
Moving forward is the key;
It's the image of good governance
And of a people who are free.[3]

RAOUL SUÉRUS, "THE VELOCIPEDE" (1869), TRANSLATED FROM LATIN BY ROGER MACFARLANE

By far the longest and most substantial poem devoted to the velocipede during its heyday, "Le Vélocipède" (The Velocipede), was penned by Raoul Achille Suérus (1850–1930) for the Saint-Charlemagne banquet in 1869. This banquet was celebrated in French schools each year on January 28, the traditional date of both Charlemagne's death and the date he was believed to have been canonized (though his canonization was never recognized by the church). At each *lycée* (or high school), a student would be selected to compose a poem for the banquet. These poems, written in French or Latin, typically included inside jokes about classmates or professors and frequently satirized the educational experience since Charlemagne was thought to be the founder of the French educational system.

Suérus, who would have been an eighteen-year-old rhetoric student at the time, went on to become a professor, and then a principal; he also authored a number of textbooks and founded a geographic society.[4] His poem, originally written in Latin in dactylic hexameter and graciously translated for this project by classicist and cyclist Roger Macfarlane (Brigham Young University), includes a number of themes covered in this book, including the velocipede as a symbol of innovation and change, as superior to the horse, as a manifestation of the carnivalesque, as mythological, and so on. While not written for wide distribution, and likely only heard and read by a very limited audience, it offers an example of how the obsession with the velocipede reached into even the smallest pockets of French culture. We include it here as a short epilogue, as a poetic summary of the spirit of velocipedomania.

This yearly celebration, alas, leads us to again repeat the same
banquet in our school,
And the same dishes appear upon the solemn table as offerings
To our ravenous bellies. But thou, our great headmaster,
Hast cautiously despised new things and kept them from our reach.
Behold: every day all things around us are changing.
Now novelty reigns; novelty's feverish passion
Drives us: we fly boldly amidst the great unknown.
Nowadays the young scholar wears a foppish uniform like
Trossulus,[5]
It is a time when a great theater is erected for Euterpe,[6]
When a girl is taught from the Sorbonnic chair,[7] when
Chassepot rifles and mobile infantry threaten the foe:[8]
Forthwith, lo, from fertile mind of man a machine—
Like panoplied Pallas from Jupiter's prodigious brain[9]—
Hath sprung: the name they give it is the *velocipede.*

A new carriage stands there (imagine it!): light wheel to light wheel
is joined;
A man sits astride it on a saddle;[10] he applies his legs
To each wheel, like a modern-day Ixion:[11] nor does any delay ensue,
but with
Alternate movement of the feet, he accelerates in swift impulse.
Look! He takes the road: thou king of birds, now, give way!
Thou also, light-footed deer; the ground hardly feels his passage,
Even the wind stands in awe: lightning's flash is not more quick.
It does not travel like a locomotive machine moved by billowing
steam,
Blithely only on its designated track: nay, it seizes
Every trodden path; as long as the rider's deft hand
Steers it, the machine obeys responsively and follows the driver.
Farewell, thou ancient carriage built of unflexing wood,
Thou that hammered our weary bones with harsh concussions!
Before these witness, and for all time, coachman, I bid you farewell!

How often have you mocked my cries as I've chased after you—
you scoundrel!
Horses, farewell! You who are now skin and bones,
Since birth you have grown accustomed to your shambling gait;
Thou, too, noble stallion, haughty light-hoof[12]—
May the same misfortunes unite you all as one. To wit,
May now that bovine lot receive you all. The earth shall ye till with
the plow,
Till wearied in your bodies and with senectitude slowing
your strength
Your limbs are handed over to the butchers' knife and
Your flesh dearly sacrificed will sate our bellies—finally a new dish!
To return to the point: things people once hoped for are now
spurned;
Every man now is himself his own coachman. Possessing feet raises
one to equestrian status.
Anybody who has this newfound quadruped hardly requires lavish
supplies of grain
For daily fodder nor immense buildings for the stables;
It stands ever ready at hand and flies forth indefatigable.
The desire to ride again is strong enough to temper yesterday's ride:[13]
Nor is there any report that
Any velocipede has rebelled against its brake or its rider and
Bolted suddenly in a raging escape into the broad plowlands.

Everyone loves the velocipede: everyone desires to own one—
The slovenly, the industrious; the serious, and the merry.
Have not some of you, professors, overcome your stodgy dignity and
Personally mounted this new horse? 'Tis no cause for shame.
C'mon you other gents, try it for a first time!
No grumpy old censor will enter our school after hours;
No one will sharply scrutinize us like sinners, standing by and
Turning his thousand watchful eyes. No one will loom over us in a
foul mood,

Rebuking us while he cracks lively jokes like these:
"What's taking so long? Your pace is pretty slow. Alas!
I could have written five hundred hexameters faster than this!"
When the awaited bells finally sound, everybody
Flees the school's wretched buildings; and when we cast off
The rigid veneer of antiquity, many of us,
Resplendent in every detail, our tresses curled and crimped,
Shall jump onto the velocipede, and explore the paths of
The teeming forest. There the dandy hoists the prizes.
This way and that he steers the wheel with masterful skill, where
The nimble deer follows the fragrant prints of the tawny doe.
Fortune denies these things as long as the sad academy locks us in.
Take heart! But look how novelty enters into this shrine now sullied with rust.
Athletic training, weapons and weights preoccupy even us.
New things—we can always hope—arise to meet us.
Perhaps there will come a day when, no longer a dungeon and cave,
Our school will smile attractive in sparkling refinement,
A garden watered with pure streams and fragrant with flowers:
Then shalt thou, machine of our dreams, pull up beside these walls.
O greatest seer, father and assiduous president,
O thou revered for such constant vigilance, such deep goodness,
We pray that, though but newly in our midst, you may earn our long-standing affection,
That you favor us now, that you come now to our aid,
And that those days we yearn for may shortly come to pass.
What joy will then bear sway! When that happy dawn will shine in seven days' time,[14]
The dawn on which we may leave the school in the long procession,
We shall carouse through the city's neighborhoods upon unweariedfoot:
A long parade will proceed gleeful upon these new horses;
No longer will they hesitate to undertake new trips far from the city.

When the throng seeks grassy fields far flung and the open
 countryside,
They will give themselves over to the diversions: many will delight in
 vying to see
Who first prevails to reach the finish line.
The youth stand ready. But who shall be first? Bounteous Jove!
He Himself appears—can I believe my eyes?—He himself presides
 o'er them,[15]
The official who keenly seeks the prize of the lovely palm.
The clarions sound, the throng rushes with equal ardor; yet,
Fortune does not favor all equally. Even now he stretches for the goal,
Exalted by hope, over-ardently he incites his mount. Lo, alas!
He yields, and his vain hope for the illustrious triumph perishes;
He yields, and the vulgar crowd mocks the woeful peer.

The passionate column flies, black dust flies into the breeze
From the turbines. But, what passion has caught me up? Whoa!
Pegasus, hold your gait; the winged race becometh not thee,
 aged steed.
Stand still, lest this throng laugh also at you in your failure.
The confident bard climbs Phoebus's wild peaks alone upon the
 Velocipede's lightning strike.[16]

CONCLUSION

"WE THOUGHT THE VELOCIPEDE WAS DEAD"

In his history of the velocipede, Keizo Kobayashi remarks that by mid-1870, competition and lawsuits between manufacturers had already begun to significantly weaken the Paris velocipede market. But, he adds, the Franco-Prussian War "struck what could be considered a fatal blow to the velocipede industry."[1] In the fall of 1870, as Prussian troops began a siege of Paris, Le Grand Jacques, for his part, suspended publication of his newspaper. When he began publishing it again on July 16, 1871 (under the new title *La Vitesse*), he reported the following:

> During the siege of Paris, the offices of the *Vélocipède illustré* received visits from people interested in organizing military troops of velocipedists to serve in the war. . . . But such troops cannot be created properly at the last minute; we hope that there will be better advance planning in the future. . . . Several individual, practical attempts were made and, in the absence of formally organized military companies, we saw dispatch riders crisscross Paris during the siege on velocipedes. It must be said, however, that they were viewed with a certain amount of consternation. The general population's mood, which had become more serious, persisted in classifying exercise on a veloce as a form of entertainment. It remains an unassailable truth that objects of the most obvious utility are the most difficult to impose on the habits of the masses.[2]

Le Grand Jacques's report strikes a melancholic chord, remarking that Parisians viewed the velocipede as entirely incompatible with life under the siege. This is undoubtedly because before the war the velocipede had been seen as a hopeful invention linked with freedom of movement, upward mobility, gender equality, and the carnivalesque. Le Grand Jacques himself, in his *Manual of the Velocipede*, had coded the velocipede as transgressive, liberating, and erotic—qualities that were out of place in a time of war and privation.

A four-part series—"Journal d'un véloceman pendant le siège de Paris" (Journal of a Velocipedist during the Siege of Paris) by Charles D . . . —published in *La Vitesse*, underscores the clash between the ethos of the velocipede and the somber mood of the siege: "My ride this morning troubled me. People no longer understand the bicycle. They were giving me angry looks, as if I was somehow insulting the nation's misfortunes. An idiot even called me a Prussian. This is insane. I'm fed up."[3] Since the velocipede was so connected with fun and frivolity, riding represented a sort of treason, a denial of the tragedy that was unfolding throughout the city.

After cleaning and storing his velocipede, the author of the velocipedist's journal describes walking through Paris in the fall of 1870: "I haven't seen a bicycle all day, although I spent a few hours strolling around the Champs-Elysées. What is Le Grand Jacques going to say? But he won't say anything, since his journal is no longer being printed. . . . My God! How far we are from the days of Velocipede races."[4] In another installment, he laments, "There is fighting near Villejuif and Châtillon. The Western telegraph wire has just been cut; it was the only one keeping us connected to the departments. We are now living in the dark."[5] The velocipede had allowed for quicker transportation, more communication, and faster, more reliable delivery of news before the war; under the siege, newspapers stopped circulation and communication with the world outside of Paris was cut off. The velocipede, the great symbol of freedom and communication in the late 1860s, no longer had a place in the confined, suffocating world of the siege and the Commune.

By August 1871, Le Grand Jacques had begun taking stock of the consequences of the siege:

> NEWS.—It should not be believed that *La Vitesse* easily regained the clientele that the former *Vélocipède illustré* had worked so hard to create. Unfortunately, the war has left its mark there as well. The very nature of our journal has made our situation more difficult than that of our colleagues. Among our readers and subscribers were bold young people who donned the uniform at the first call of the Fatherland. The least fortunate defended our ramparts and ate the gray bread of the siege. We are sorely aware of this.—In the last few weeks, a certain number of issues that we sent to subscribers were returned to us, with the annotation "DEAD" written next to the address.—To these lost friends, all our condolences.
>
> A greater number of copies have been returned to us with these words: *Unknown,—Left, no known forwarding address,—Absent.* . . . This gives an idea of the upheaval of our beloved country.[6]

Before the war, *Le Vélocipède illustré* was published twice weekly. Afterward, it was published only six times as *La Vitesse* in July and August 1871, then once a week from May to October 1872, and then not again until 1890.

In late spring 1872, as Le Grand Jacques was relaunching his newspaper under its previous title, *Le Vélocipède illustré*, he summed up the *Année terrible* in these terms: "It's not that the Velocipede did not have to endure great trials in the grim times through which we have just passed. But it was the common fate, and the Velocipede suffered all the more because its followers and supporters were among the country's agile and militant youth. Our ranks suffered tremendous losses, and after a fatal struggle, we thought the Velocipede was dead. Like the phoenix, it rises from its ashes."[7] *Vélocipédie*, like France itself, suffered during the siege and the Commune. Another spring gave hope to a weakened population, a struggling industry, and a pastime that had seemed out of place amid the gravity of war. The furor the velocipede had inspired never returned to prewar levels. It would take new technologies like chains, gears, and pneumatic tires for the mania of the two-wheeler to fully reignite in the late 1880s. In 1867–69, the velocipede was presented as the embodiment of liberty, liberation, mobility, and useful technology. During the Siege of Paris and the subsequent Commune, the velocipede could offer little more than a frivolous use of wheels at a time when cannon wheels were needed and when technology was subverted and used for bellicose, destructive, and oppressive ends.

Other manuals on the velocipede were published after the war and the Commune. In 1872, Rémy Lamon published *Théorie vélocipédique et pratique, ou manière d'apprendre le vélocipède sans professeur* (Velocipedic Theory and Practice or How to Learn the Velocipede without a Teacher). As the title page explains, Lamon was a lieutenant in the reserves (a conscripted branch that served during the Franco-Prussian War) and had finished fourteenth in the Paris-Rouen race of 1869, a race incidentally organized by *Le Vélocipède illustré*. Lamon acknowledges that velocipedia "had developed a great deal before the war but was stopped by the events of 1870 and 1871 that were so deadly for our troops. Let us now hope that the velocipede will move forward stronger than before."[8] Lacking the quirkiness of de la Rue's *Note* and the eclectic fancy of Lesclide's *Manual*, Lamon's manual offers a comprehensive, step-by-step guide to riding a velocipede: how to mount, how to find the pedals, how to brake, how to stand up while riding, and so on. Lamon even pours cold water on velocipedic enthusiasm, writing, "The velocipede is an innocent exercise that prevents chimeras or cloud castles from filling one's mind; it gives strength and courage to your body and doesn't leave you time to do things that might be harmful; it gives birth to the most agreeable thoughts, and allows you to emulate your daily work: the state you have chosen or to which you are destined."[9] In other words, Lamon sees riding the velocipede as an extension of the capitalist workplace, as a means to avoid flights of fancy and imagination, as a way to engage one's body in utilitarian ways. Lesclide's *Manual*, on the contrary, encouraged readers to pedal down side alleys, to embrace the transgressive, to mix "flights of fancy" with practical advice; it alternated between fiction and information, between eroticism and pragmatic benefits to the body. To be fair to

Lamon, in 1872 the velocipede was no longer new and could never recapture the magic of the late 1860s. The promise of social change and of artistic and technological creation drove velocipedomania, but, in the aftermath of the war, technology was viewed more as a means of destruction than of creation.

Nevertheless, the cultural forms of velocipedomania sowed important seeds in the French social imaginary; while they would remain dormant until new resources, designs, and production methods led to the development of the modern bicycle, the early promises of velocipedomania familiarized the French population with new ways of thinking about, and participating in, notions of speed and freedom. By offering the promise of mobility across genders and classes, the velocipede gave important momentum to its fin de siècle successor.[10] Consider an 1890 poster for Cycles Humber by the illustrator Alfred Choubrac (figure C.1). It depicts a woman on a safety bicycle with inflatable tires and a brake lever on the handlebars. While the bike looks very similar to machines seen on city streets in the twenty-first century, the overall composition closely resembles illustrations of women on velocipedes from the late 1860s. Like the velocipedist in figure I.5, this woman is attired in red and yellow, and her bicycle is aligned at the same angle as the velocipede in Hadol's earlier sketch. Her flowing cape-like flourish is reminiscent of the velocipedist's trailing tails. Additionally, as with the velocipedist, the contours of this cyclist's left leg are fully visible as they push down on the pedal. This 1890 advertisement also compares with Benassit's illustration of a young woman riding a velocipede (figure 3.27). The bicycles are again aligned in the same direction, both women's lower legs are exposed, and flowing dresses reveal bare arms. Finally, though the 1890 cyclist's cap is adorned with a star-like decoration, it is not dissimilar to the feathers worn by the female velocipedists in figures I.5 and 3.20. In short, as with other tropes associated with the bicycle, its fin de siècle visual representations build on earlier images of the velocipede.

Although velocipedomania hit the proverbial wall in 1870, its primary proponents kept their wheels turning: Lesclide, Paz, and

FIGURE C.1. Alfred Choubrac, advertisement for Cycles Humber, 1890.

FIGURE C.2. Abel Amiaux, cover of *Pédalons*, 1892.

Blondeau continued writing about two-wheelers well into the 1890s. In 1890, at the age of sixty-five, Lesclide married his second wife, twenty-four-year-old Mélanie Ignard (1866–1951; see chapter 3). Under her married name Juana Richard-Lesclide—and the pen name "Jean de Champeaux"—she was as fervent a supporter of the two-wheeler as Lesclide. In addition to continuing as editor of *Le Vélocipède illustré* after her husband's death in 1892, Jean de Champeaux became one of the founders of the Touring club de France.[11]

Just before he died, Lesclide wrote a letter praising a collection of stories by Jehan de la Pédale (pseudonym of the journalist Pierre Lafitte). The letter features prominently in the front matter of *Pédalons!*, a book that focuses on the velocipede's descendant, the bicycle (figure C.2). Lesclide expresses regret for the loss of his youth while applauding the stories in *Pédalons*: "They are young, lively, gritty, emotional, and the reader's heart beats in harmony with the impulsive bicycle that links them."[12] Lesclide's description of these "young, lively, gritty, emotional" stories could have equally been applied to the stories inspired by the velocipede that he had written a quarter century earlier. This letter can also be read as a symbolic passing of the torch from the velocipede era of Lesclide's *Manual of the Velocipede* to the bicycle craze of the 1890s. Finally, it reminds readers that the enthusiasm for the bicycle, an enthusiasm that would lead to the creation of the Tour de France in 1903, was formed and culturally embedded in France decades earlier.

ACKNOWLEDGMENTS

We thank the College of Humanities at Brigham Young University (BYU) for their financial and logistical support; in particular we thank Dean Scott Miller and secretary to the dean Shasta Hamilton. Maggie Marchant at BYU's Harold B. Lee Library provided invaluable assistance in finding and reproducing high-quality images from the original *Manuel du vélocipède*. We also thank the professional and student editors at BYU's Faculty Publishing Service for their help with the manuscript. Roger White, curator of the Smithsonian Institution's National Museum of American History, kindly granted us access to some of the velocipedes in its collection. We can neither confirm nor deny reports that we took a few for a spin. Thanks to Racer Gibson for keeping Corry's bicycle in good repair, allowing him to keep his passion for the two-wheeler in peak form. Finally, we are very appreciative of Bucknell University Press: director Suzanne Guiod offered enthusiastic support for this project, managing editor Pamelia Dailey provided tireless editorial help, and the hard work of the entire team at the press kept this book balanced and gliding effortlessly forward.

NOTES

INTRODUCTION

Epigraph: Charles Yriarte, "Courrier de Paris," *Le Monde illustré*, June 26, 1869, 402. Note on translated text: Unless otherwise stated, all translations are our own.

1. We will discuss the impact of the Siege of Paris and the Commune on the velocipede in the conclusion. Given the "troubles" in France, during the 1870s and 1880s the velocipede evolved primarily in Britain where the high bicycle and then the bicyclette would be developed. Our contention, however, is that the cultural imprint of velocipedomania paved the way for the eventual flourishing of the modern bicycle in France.

2. *Note sur le vélocipède à pédales et à frein de M. Michaux par un amateur* (Paris: Imprimerie de Ad. Laine et J. Havard, 1868); Henri Blondeau, *Dagobert et son vélocipède*, lyrics by Frédéric Demarquette (Paris: Ch. Grou, 1869); Le Grand Jacques, *Manuel du vélocipède* (Paris: Librairie du Petit Journal, 1869). In the notes, titles of these works will remain in French since the page numbers are from the original French publications included in the bibliography.

3. Throughout this book, we use the terms *imaginary* and *social imaginary* as they have been used by social philosophers and cultural historians (e.g., Cornelius Castoriadis, Alain Corbin, Dominique Kalifa) to mean the shared network of meaning established through recourse to all cultural artifacts (high and low, major and trivial) in a given culture at a particular time.

4. It should be noted that the term *bicycle*, in the generic sense of a two-wheeled vehicle, had been in use in the 1820s. An 1828 edition of the *Journal des artistes* attests that a two-wheeled horse-drawn "cabriolet" was also called a "Bicycle" (C. V., "Lettre d'un convalescent, sur les travaux et embellissemens [*sic*] de Paris," *Journal des artistes*, September 7, 1828, 148). And the term velocipede was used to refer to the draisine and after 1870 to refer to high-wheeled vehicles (also called *grand-bi* in French). For our purposes, we use the word *velocipede* to mean the wood and iron vehicle conceived of by Michaux in the 1860s and informally introduced at the Exposition universelle in Paris in 1867. We use the term *bicycle* to refer generally to the modern or "safety" bicycle.

5. See Tony Hadland and Hans-Erhard Lessing, *Bicycle Design: An Illustrated History* (Cambridge, MA: MIT Press, 2014), 40–53.

6. For a discussion of Chapus and his connection to French sport, see Corry Cropper, *Playing at Monarchy* (Lincoln: University of Nebraska Press, 2008), most notably the introduction and pp. 104–15.

7. G. d'O, "En véloce! En véloce!" *Le Sport, Journal des gens du monde*, July 28, 1867, 3.

8. G. d'O, "En véloce! En véloce!" 3.

9. G. d'O, "En véloce! En véloce!" 3.

10. For more on the nexus of French identity, universalism, and its geographic space, see Dana Kristofor Lindaman, *Becoming French: Mapping the Geographies of French Identity, 1871–1914* (Evanston, IL: Northwestern University Press, 2016).

11. Thomas Burr, "Markets as Producers and Consumers: The French and United States National Bicycle Markets, 1875–1910" (PhD diss., University of California–Davis, 2005), 67–68. See also Alex Poyer, "The Origins of Velocipede Clubs," in *Le Vélocipède: Objet de modernité*, ed. Nadine Besse and Ann Henry (Saint-Etienne: Musée d'art et d'industrie, 2008).

12. For a short history of Karl von Drais's two-wheeled running machine and its evolution, see David Herlihy, *Bicycle: The History* (New Haven: Yale University Press, 2004), 15–52; and Hans-Erhard Lessing, "What Led to the Invention of the Early Bicycle?" in *Cycle History 11: Proceedings of the 11th International Cycling History Conference*, ed. Andrew Ritchie and Rob van der Plas (San Francisco: Van der Plas Publications, 2001). For his part, Smethurst is skeptical of the connection between the draisine and the velocipede (*The Bicycle: Towards a Global History* [London: Palgrave Macmillan, 2015], 23–28). The French spelling, *draisienne*, is frequently used; however, we follow the guidance of Hadland and Lessing, who recommend the spelling *draisine* in the preface to their book *Bicycle Design*.

13. As Vanessa Schwartz maintains in her book *Spectacular Realities: Early Mass Culture in Fin-de-Siècle Paris* (Berkeley: University of California Press, 1998), "Paris had enormous power to 'represent.' When it came to 'modernity,' Paris stood for things French" (6). In other words, what happened in Paris had broad meaning for France and modernity as a whole. Notably, Michaux's velocipede was discussed and advertised in newspapers all over France (see, for example, *Le Petit Marseillais*, January 25, 1869; *Le Courrier de Bourges*, July 16, 1869; *Courrier de Saône-et-Loire*, July 27, 1869; *L'Indépendant de la Charente-Inférieure*, August 28, 1869). Advertisements in non-Parisian papers included the note, "Delivery available in rural areas." Though we focus on Parisian representations of the velocipede in this book, the depictions undoubtedly had resonance throughout the country and Parisian promoters attempted to align the velocipede with a certain national character.

14. Léon Bienvenu, "Fantaisie," *L'Eclipse*, February 16, 1868, 3.

15. Léon de Villette, "Nouvelles locales et faits divers," *L'Industriel de Saint-Germain-en-Laye*, May 9, 1868, 1.

16. Le Grand Jacques, *Manuel du vélocipède*, 6.

17. Timothée Trimm, "Les Vélocipèdes," *Le Petit Journal*, July 5, 1868, 1.

18. Trimm, "Les Vélocipèdes," 1.

19. Trimm, "Les Vélocipèdes," 1.

20. The first volume of the well-known poetry collection *Le Parnasse contemporain* (The Contemporary Parnassus) had come out in 1866 and featured poetry by two of the poets on Trimm's list: Théodore de Banville and François Coppée. In the very same month as this article by Trimm was published, Banville wrote a poem on the velocipede, "L'Homme vélocipède" (discussed and translated in chapter 4 as "Here Comes Velocipede Man").

21. Trimm, "Les Vélocipèdes," 1–2.

22. Bienvenu, "Fantaisie," 1868, 3.

23. Alexis-Georges Favre, *Le Vélocipède, sa structure, ses accessoires indispensables, le moyen d'apprendre à s'en servir en une heure* (Marseille: Barlatier-Feissat père et fils, 1868), 38.

24. Favre, *Le Vélocipède*, 22–23, 36. It is worth noting that Favre (on pp. 29 and 31) refers to the *Note on Monsieur Michaux's Velocipede*—translated in this book—as a model for promoting the practical uses of the velocipede.

25. Élie Bellencontre, *Hygiène du vélocipède* (Paris: L. Richard, 1869), v.

26. Alfred Berruyer, *Le Manuel du véloceman* (Grenoble: Prudhomme, 1869), 5. In addition to his significant architectural work in Grenoble, Berruyer also patented the first kickstand for a velocipede and came up with the idea for a bike path. On Berruyer's architectural work, see Cédric Avenier, "Alfred Berruyer (1819–1901): La Volonté d'un architecte diocésain," *La Pierre et l'écrit* 13 (2002). On his invention of the kickstand, see Alfred Berruyer, *Caducêtres ou jambes étrières brevetées s. g. d. g. pour vélocipèdes* (Grenoble: F. Allier père & fils, 1869). Finally, Nicolas Grenier maintains that Berruyer is the "inventor of the bike path" (*La Petite Reine: Une anthologie littéraire du cyclisme* [Le Crest: Les Editions du Volcan, 2017]), 98.

27. For more on the rumors of war with Prussia in September 1868, see the article below Daumier's illustration in *Le Charivari* (Henry Fouquier, "Courrier de Paris," September 17, 1868) and René Rivarol's article "Entre deux bourses," *Le Figaro*, September 16, 1868.

28. We have recorded Baron's quadrille along with other music inspired by the velocipede and made the recordings available online at http://velocipede.byu.edu.

29. Alexandre Six, "Les Vélocipèdes" (Lille: Imprimerie de Six Horcmans, 1869).

30. Blaquière wrote the song for Henri Thiéry's operetta *Les Contributions indirectes* (Paris: Dentu, 1866). It premiered at the Théâtre des Variétés in 1865.

31. Alexandre Flan, "Les Vélocipèdes," *La Chanson illustrée*, May 2, 1869, 2. The Pont Neuf bridge in Paris features a well-known equestrian statue of the French king Henri IV seated on his horse. The statue was completed in 1618, shortly after the king's death, and rebuilt in 1818 after it was destroyed in the French Revolution. Figure 3.18 also humorously depicts a velocipedist memorialized in sculpture.

32. Le Grand Jacques, *Manuel du vélocipède*, 94.

33. See chapter 3 of the *Manual of the Velocipede* by Eugène Paz, "On the Topic of Exercise."

34. As Rosemary Lloyd notes about fin de siècle bicycling posters, "The freedom depicted as coming to women through cycling was far less than one might have expected"; the same is no less true for the bicycle's predecessor, the velocipede. The lived reality rarely met the promise of freedom for women. Rosemary Lloyd, "Reinventing Pegasus: Bicycles and the Fin-de-Siècle Imagination," *Dix-Neuf* 4, no. 1 (2005): 53.

35. See Eugen Weber, *France: Fin de Siècle* (Cambridge, MA: Harvard University Press, 1986), 37. First published in 1800, an ordinance from the Paris Prefecture of Police prohibited women from wearing pants unless they received written authorization from the mayor or the police. See Christine Bard, *Une histoire politique du pantalon* (Paris: Editions du Seuil, 2010), 79–80. In this political history of pants, Bard contends that sports—primarily cycling—led to the nineteenth-century's most significant changes to female sartorial norms (192–205).

36. Clairville (pseudonym of Louis-François-Marie Nicolaïe) and Paul Siraudin, *Le Mot de la fin: Petite revue en un acte et deux tableaux* (Paris: Michel Lévy Frères, 1869).

37. Edgard Pourielle, "Théâtre de l'Athénée, *Le Petit Poucet*," *Le Théâtre illustré*, October 1868, 2.

38. "Nouvelles des théâtres," *Théâtre-Journal: Musique, littérature, beaux-arts*, July 5, 1868, 3.

39. Arthur Arnould, "Vélocipédons," *Le Journal amusant*, June 13, 1868, 2.

40. Le Cousin Jacques, "A vélocipède! Messieurs, à vélocipède!" *L'Eclipse*, May 16, 1869, 2.

41. Léon Bienvenu, "Fantaisie," *L'Eclipse*, August 22, 1869, 2.

42. "Vélocipède à pédales du prince impérial," Napoleon.org, https://www.napoleon.org/histoire-des-2-empires/objets/velocipede-a-pedales-du-prince-imperial.

43. The first reference to the imperial prince as Velocipede IV appears to be in an article by Le Numéro 445 (typical of the pseudonyms used in *Le Figaro*) titled "Courrier de pélagie," *Le Figaro*, June 8, 1870, 1; it is repeated in the same newspaper two days later (June 10, 1870, 1). Le Grand Jacques's newspaper, *La Vitesse*, lamented the sobriquet, begging its readers, "for the sake of the velocipede," not to call him Velocipede IV, adding, "The velocipede is essentially democratic" ("Faits divers," July 23, 1871, 3).

44. Le Grand Jacques, *Manuel du vélocipède*, 6.

45. Jean-Marie Brohm, *Sport: A Prison of Measured Time*, trans. Ian Fraser (London: Ink Links, 1978), 5.

46. Smethurst, *The Bicycle*, 31.

47. Philip Nord, *The Republican Moment: Struggles for Democracy in Nineteenth-Century France* (Cambridge, MA: Harvard University Press, 1995), 191.

48. Alan R. H. Baker, *Amateur Musical Societies and Sports Clubs in Provincial France, 1848–1914: Harmony and Hostility* (London: Palgrave Macmillan, 2017), 7.

49. Baker, *Amateur Musical Societies and Sports Clubs*, 171–72.

50. After France's military defeat to the Prussians in 1870, gymnastics would become synonymous with military exercises and shooting was often part of the regimen; in fact, gymnastics would become required in schools beginning in 1880. See Pierre Arnaud, *Le Militaire, l'écolier, le gymnaste: Naissance de l'éducation physique en France (1869–1889)* (Lyon: Presses universitaires de Lyon, 1991).

CHAPTER ONE THE UTILITARIAN VELOCIPEDE

1. Monsieur de la Rue, "L'Insubmersible: Appareil de sauvetage, de sport, d'hygiène, de plaisir et d'utilités diverses," *Le Monde illustré*, March 27, 1869, 197–98.

2. Elected to the National Assembly in 1869, Léon Gambetta emerged as a leader of the opposition to the empire and declared the creation of the Third Republic after the fall of Napoleon III.

3. Louis de Coulanges, "Les Préfets de la république," *Le Figaro*, March 9, 1872, 1.

4. Keizo Kobayashi (*Histoire du vélocipède de Drais à Michaux, 1817–1870: Mythes et réalités* [Tokyo: Bicycle Culture Centre, 1990], 215) believes the *Note*'s author was Aimé Olivier, one of the owners of Michaux's company, while Scotford Lawrence indicates in the preface to his translation of the *Note* that it was Georges de la Bouglise, a friend of the Olivier brothers (see *The Velocipede: Three Contemporary French Texts*, trans. Scotford Lawrence [Cheltenham: John Pinkerton Memorial Publishing Fund, 2014], 4). Unfortunately, neither study provides any evidence. Given the connection between de la Rue, the Ministry of the Navy, the nautical velocipede, and his acknowledgment of a recent publication on the velocipede, we remain confident that the Baron de la Rue was the author of this text.

5. *Note sur le vélocipède à pédales et à frein de M. Michaux par un amateur*, unnumbered page.

6. "Note sur le vélocipède," *Le Dartagnan*, no. 45, May 16, 1868, 4. In this newspaper, the "Note" is preceded by an announcement for the Paris véloce-club, which is followed by text about the Saint-Cloud velocipede race and then by an advertisement for the upcoming music series at the Pré Catelan (a series referenced in the *Manual of the Velocipede*).

7. Road locomotives were steam-powered tractors, primarily used on farms.

8. The "celerifere" was a type of public horse-drawn carriage patented by Jean-Henri Sievrac in 1817. De la Rue here conflates the definition of the celerifere—which Bescherelle's dictionary defines as a "public carriage"—with the definition Bescherelle gives for the velocipede. It appears that de la Rue did this primarily to assure a French ancestor for the velocipede. Le Grand Jacques would repeat this mistake in the *Manual of the Velocipede* a year later. Louis Baudry de Saunier would then cement the story of the celerifere in his book *Histoire générale de la vélocipédie* (Paris: P. Ollendorff, 1891), maintaining that a nobleman named de Sivrac created it in 1790 and that the celerifere, rather than the German draisine, was the velocipede's true forerunner. The eighteenth-century noble de Sivrac was as fictional as the two-wheeled celerifere itself. See Kobayashi, *Histoire du vélocipède*, 327–31.

9. The front wheel of the velocipede is typically larger, but the brakes are uniformly affixed to the rear wheel. See the illustrations in the *Manual of the Velocipede* in chapter 3.

10. The Ministry of the Navy was located at the place de la Concorde, about 3.9 kilometers from the Bastille on the rue de Rivoli. The rue de Rivoli is still a congested road that runs next to the Louvre and the Tuileries in the heart of Paris.

11. The boulevard de Strasbourg runs from the gare de l'Est in Paris and changes into the boulevard de Sébastopol, which continues down to the Seine, a route of about 2.4 kilometers.

12. Noisy-le-Grand and Noisy-le-Sec are both towns outside of Paris. The author is most likely referring to Noisy-le-Grand since there is a fairly steep descent from the center of town to the Marne River below.

13. Hippolyte Triat and Eugène Paz both ran large exercise facilities in Paris in the nineteenth century. Eugène Paz's gymnasium on the rue des Martyrs, the same street as Lesclide's office, can be seen in figure 1.3, a lithography created by Victor Rose around 1866. Paz also contributed to Le Grand Jacques's *Manual of the Velocipede*. See the chapter of the *Manual* titled "On the Topic of Exercise."

14. Cochinchina is a historical exonym for a southern region of Vietnam, also known as Quinam. The area was a French colony from 1862 to 1954.

15. St. Hubert is the patron saint of hunters.

16. The daily newspaper *La Patrie* was closely aligned with the empire of Napoleon III. A short dispatch by E. Bouchery on the front page on April 7, 1868, notes that appointments are being considered to oversee the improvement of rural roadways.

17. After serving as prefect in the Nièvre department, Henri Michon, vicomte de Vougy (1806–91), was appointed director general of telegraph lines in 1853 and served in this position through the 1860s. See Alexis Belloc, *La Télégraphie historique: Depuis les temps les plus reculés jusqu'à nos jours*, 2nd ed. (Paris: Frimin-Didot, 1894), 221; and Pierre et Paul, "Notes et souvenirs," in *Le Roannais illustré*, series 6 (Roanne: n.p., 1892), 34.

18. The Société centrale de sauvetage des naufragés was founded by imperial decree on February 12, 1865, and placed under the leadership of Admiral Rigault de Genouilly. Rigault de Genouilly would become Minister of the Navy from 1867 to 1870, at the very time de la Rue was writing this manual and working at the same ministry. See François Bellec, *Les Sauveteurs: Histoire folle et raisonnée du sauvetage en mer* (Douarnenez: Chasse-Marée, 2008), 38.

19. *Galignani's Messenger* was an English-language daily paper published in Paris by an Italian publisher.

20. The sovereign is Louis-Napoleon Bonaparte, Napoleon III, emperor from 1852 to 1870.

CHAPTER TWO THE VELOCIPEDE ON STAGE

1. Playwright and novelist Albert Delpit authored an analysis of theater regulations in which he discusses the consequences of the 1864 decree at length. See "La Liberté des théâtres," *Revue des deux mondes* 25 (1878). Deregulation, an important political philosophy of the late Second Empire, meant that theatrical

genres could be mixed: acrobatic feats (like riding a velocipede) had previously been confined to the Cirque olympique on the boulevard du Temple or to hippodromes but could now be incorporated into musical productions and plays. See John McCormick, *Popular Theaters of Nineteenth-Century France* (New York: Routledge, 1993), 28–44.

2. Offenbach's most famous musical send-up might be his piece "Les Oiseaux dans la charmille" from his work *Les Contes d'Hoffmann*; it musically resembles the famous aria from Mozart's *The Magic Flute*.

3. *Nos enfants*, a drama by Ernest Rasetti, premiered at the Théâtre de la Gaîté on September 23, 1868. A review of the play by Adolphe Stel in *L'Indépendance dramatique* (September 30, 1868, 1–2) dismisses the work as clichéd melodrama and indicates that Mr. Gaillard played the role of Isidor Pavart, a criminal accomplice.

4. The operetta *Les Croqueuses de pommes* premiered September 29, 1868. Since velocipedes are not mentioned in the script, they were likely employed between acts or as part of a tableau. An article by X. in *Le Gaulois* ("Petites nouvelles," September 21, 1868, 3) indicates that the velocipedes were ridden by twelve actresses dressed as women from the rural region of Normandy.

5. P. de Faulquemont, "Chronique de Paris," *L'Indépendance dramatique*, September 23, 1868, 1.

6. A. Vautier, "Chronique," *L'Indépendance parisienne*, October 16, 1868, 2.

7. Benedict, "Chronique musicale," *Le Figaro*, October 11, 1868, 2.

8. E. Leterrier, A. Vanloo, and Laurent de Rillé, *Le Petit Poucet: Partition piano et chant* (Paris: Colombier, 1868), 125–29. A recording can be found online at http://velocipede.byu.edu.

9. Julien Deschamps, "Théâtre Molière," *Théâtre Journal: Musique, littérature, beaux-arts*, January 10, 1869, 1.

10. Deschamps, "Théâtre Molière," 2.

11. Other reviews of *Paris-Vélocipède* include P. de Faulquemont, "A Travers Paris," *L'Indépendance dramatique*, November 25, 1868, 2; and Théophile Deschamps, "Echos de Paris et des théâtres," *L'Indépendance dramatique*, January 6, 1869, 2.

12. Max Sacerdot, "Menus-Plaisirs," *Le Théâtre illustré*, January 1869, 2.

13. Rouvrelle, "Menus-Plaisirs," *La Comédie*, January 3, 1869, 4.

14. Maxime Aubray, "Vélocipède & Hippophagie," *Le Petit Marseillais*, January 8, 1869, 1.

15. Clairville and Paul Siraudin, *Le Mot de la fin* (Paris: Michel Lévy Frères, 1869).

16. A. Benezech, "Revue des théâtres," *L'Indépendant français*, January 31, 1869, 3. Benezech adapts a line from the final scene of Jean Racine's play *Esther* (1689): "Je n'ai fait que passer, il n'était déjà plus."

17. Advertisements in newspapers indicate that the theater was located at 14, boulevard Richard Lenoir. See, for example, the advertisement in *L'Orchestre: Revue quotidienne du théâtre*, September 13, 1869, 4. The space—with seating for 250 spectators—opened in 1865 as the Petit-Théâtre, then changed to the Folies Sant-Antoine in 1866, before rebranding as the Petits-Bouffes Saint-Antoine in

the fall of 1868. It closed in December 1873. See Nicole Wils, *Dictionnaire des théâtres parisiens au XIXe siècle* (Paris: Aux Amateurs de Livre, 1989), 155–56. The location now houses a contemporary flooring retail shop.

18. Paul Ferry, "Petits-Bouffes Saint-Antoine," *La Comédie*, no. 299, September 6, 1868, 2.

19. Ferry, "Petits-Bouffes."

20. Paul Ferry, "Théâtres de Paris," *La Comédie*, nos. 296–97, August 16, 1868, 3.

21. Jules Prével, "Petit courier des théâtres," *Le Figaro*, September 3, 1868, 3.

22. Marie-Nicolas Bouillet, *Dictionnaire universel d'histoire et de géographie* (Paris: Hachette, 1863), 454. Thirty-four editions of the encyclopedia were published in France between 1842 and 1914.

23. *Le Bon Roi Dagobert: Chanson ancienne* (Paris: A. Huré, 1863).

24. In the newspaper *La Presse* ("Théâtres," November 22, 1869), *Dagobert and His Velocipede* is advertised as still being performed in November. The operetta and the piano score were published in Paris by the Librairie Musicale et Chansonnière, editor Ch. Grou, in 1869. A quick search of theater listings in 1869 reveals that the operetta was performed in at least two venues that year: Salle de la Tour d'Auvergne (June) and Théâtre des Jeunes Artistes (fall).

25. Blondeau and Demarquette, *Dagobert et son vélocipède*, 36.

26. The velocipede races held in the imperial park at Saint-Cloud are confirmed in "Paris," *Le Petit Journal*, May 31, 1868, 3.

27. Prével, "Petit Courier des théâtres," 3.

28. While the imperial prince would only later be dubbed "Velocipede IV," by the time the operetta was written he was already associated with the velocipede. As early as October 1867, *Le Petit Journal* (among other outlets) reported that the prince rode a three-wheeled velocipede in the Tuileries ("Paris," October 30, 1867, 2). By 1868, riding the velocipede had become part of the prince's daily routine: after studying Latin and algebra, "he enjoys exercising by riding a velocipede" (Jules Moureau, "Nouvelles diverses," *Journal de la ville de Saint Quentin*, June 19, 1868, 3). In a description of the prince in the *Trombinoscope* in 1873, Touchatout (pseud. of Léon-Charles Bienvenu) writes, "Velocipede IV (Napoléon-Eugène-Louis-Jean-Joseph), Imperial Prince, more commonly known as*. . . . The young phenom went from triumph to triumph all of the same order, until the day his father realized that all he needed to complete his imperial education was to learn to ride the velocipede. It goes without saying that his learned this noble art as easily as one falls off a log" ("Vélocipède IV," *Le Trombinoscope*, no. 111, 1873).

29. Hector Monréal and Henri Blondeau, *Frou-Frou* (Paris: Albert Petit, 1898).

30. According to Kobayashi (*Histoire du vélocipède*, 257), Blondeau's dispatches appeared in *Le Vélocipède illustré* on June 10, July 4, August 1, August 12, September 5, and September 19, 1869.

31. "Course de fonds de Paris à Rouen," *Le Vélocipède illustré*, October 7, 1869, 2.

32. Blondeau and Demarquette, *Dagobert et son vélocipède*, Paris, Ch. Grou, 1869, https://gallica.bnf.fr/ark:/12148/bpt6k11839227/.

33. Riga was the stage name of Jean-Baptiste Marcelin (1850–1903). If this was indeed him, he would have been only eighteen playing the role of the fifty-year-old Eligius. Riga went on to perform at a number of important venues, most notably the Folies-Dramatiques and Gaîté theaters. We suspect that the printers erroneously switched the roles of Riga and Deberg.

34. The character's French name is Clampin, meaning a loafer or idler.

35. Deberg performed in other productions at the Petits-Bouffes Saint-Antoine and was praised glowingly by Barbey d'Aurevilly in his writings on contemporary theater (*Le Théâtre contemporain*, vol. 3 [Paris: Maison Quantin, 1889], 279–80).

36. The character's French name is Coq-en-zinc, literally a zinc rooster, the traditional form of weathervanes used on farms. The review of the play in *Le Figaro* claims that in the original production a character wore a weathervane on his head (Prével, "Petit courier des théâtres," 3). Weathervane's name points to his obsequious nature, his willingness to change his mind and follow the wind.

37. In the original, Weathervane puns on the words *qu'on fit dans* and *confident.*

38. Parisian operettas frequently sent characters to—or set scenes in—Parisian suburbs like Chatou, Nanterre, or Maisons-Laffitte, thereby adding a bit of middle-class realism and humor to the scripts.

39. A *piqueuse de bottines* used a sewing machine to assemble ankle boots. This is yet another example of an anachronism in the script. Roche-Trompette is not a real place, but Jacques Offenbach's opéra bouffe, *Le Château à Toto*, which premiered in May 1868, featured a character named Hector de la Roche-Trompette.

40. Dagobert here requests his "Bouillet." The *Dictionnaire universel d'histoire et de géographie* by Marie-Nicolas Bouillet was a well-known and widely distributed reference work. See note 22.

41. The entry on Dagobert is translated in the introduction to this chapter.

42. Located in the Bois de Boulogne in Paris's sixteenth arrondissement, the Armenonville Pavilion was built in the seventeenth century as a hunting lodge and converted into a restaurant in the early nineteenth century. Its popularity is attested in literature and the press throughout the nineteenth century.

43. The Compagnie du chemin de fer de Mulhouse was founded in 1837. See Joseph Duplessy, *Le Guide indispensable des voyageurs sur les chemins de fer de l'Alsace* (Strasbourg: V. Levrault, 1842).

44. In the original, Dagobert puns on the word *aisance*, a word that means both "luxury" and "water-closet," and on the word *celle*, a pronoun for "this" but also a homophone for *selle*, meaning "stool." It is an overwrought scatological pun.

45. In other words, Cupid.

46. A dark lantern has sliding panels that can be closed or opened to hide or reveal the light.

47. Saint-Denis, third-century bishop of Paris, is the patron saint of the French people and a patron against frenzy, strife, and headaches. Dagobert founded

the Saint-Denis basilica, where he was eventually buried alongside Saint-Denis's tomb.

48. The Chalet de la Porte Jaune was a restaurant and hunting lodge built by Napoleon III in the Bois de Vincennes. It is still in use as a conference center. It was a popular meeting and dining spot in the late nineteenth century and was even the subject of a popular song by Jules de Rieux and Charles d'Orvict titled "Au restaurant de la Porte Jaune" (Paris: Le Bailly, 1878). It was known by Parisians as a rustic and "country" restaurant (see Larousse's *Grand dictionnaire universel du XIXe siècle*, vol. 13, s.v. "restaurant" [Paris: 1875], 1050).

49. The Tyrolienne is a folk song, something like yodeling, performed here in harmony by all three characters.

50. In the nineteenth century, champagne producers began wrapping foil around the necks of bottles: silver foil typically meant it was dry wine. See Frédérique Crestin-Billet, *Veuve Clicquot: La Grande Dame de la Champagne*, trans. Carole Fahy (Grenoble: Glénat, 1992), 139.

51. The original substitutes the word *Clicot* for champagne. Veuve Clicquot, a large champagne house based in Reims, was founded in 1772 (Crestin-Billet, *Veuve Clicquot*, 56). It appears (misspelled) with some regularity in French operettas of the period; see, for example, Louis Leroy and Alfred Delacour, *Les Mormons à Paris* (1874), act 3, scene 7; Eugène Labiche and Édouard Martin, *La Poudre aux yeux* (1861), act 2, scene 8; Eugène Labiche and Alfred Delacour, *Les Petits Oiseaux* (1862), act 2, scene 2.

52. This is a clear reference to the popular Revolution-era song mentioned earlier.

53. In French, the engraving reads "Mon fils a des cheveux" (My son has hair). Goldbrick explains that it means "Mon fils a des cheveux roux/roue." *Red* and *wheel* are homophones in French. Goldbrick then exposes his red hair to prove he is Caribert's son. Thanks to Brigham Young University student Tyler Orr for coming up with the translation of this pun.

CHAPTER THREE NARRATING VELOCIPEDOMANIA

1. See Richard Sieburth, "Same Difference: The French Physiologies, 1840–1842," in *Notebooks in Cultural Analysis: An Annual Review*, edited by Norman F. Cantor (Durham, NC: Duke University Press, 1984), 162–63; and Margaret Cohen, "Panoramic Literature and the Invention of Everyday Genres," in *Cinema and the Invention of Modern Life*, edited by Leo Charney and Barbara Schwartz (Berkeley: University of California Press, 1995), 232.

2. When quoting the original and throughout our translation of the *Manual of the Velocipede* we capitalize *velocipede*, following the source text. In our analyses and in our image captions, however, we write *velocipede* in lowercase.

3. For examples of the way the bicycle embodies French cultural values, see Christopher S. Thompson, *The Tour de France: A Cultural History* (Berkeley: University of California Press, 2008); Hugh Dauncey and Geoff Hare, *The Tour de France, 1903–2003* (London: Frank Cass, 2003); and Hugh Dauncey, *French Cycling: A Social and Cultural History* (Liverpool: Liverpool University Press, 2012).

4. Lesclide writes here under the pseudonym he used frequently throughout his career: Le Grand Jacques. Many chapters in the *Manual of the Velocipede* are signed by other names and initials (often humorous plays on words), but given the unity of style and vocabulary throughout, we are fairly confident Lesclide authored many—probably most—of the chapters.

5. Jacques Seray, *Richard Lesclide: Du Vélocipède illustré à la table de Victor Hugo* (Vélizy: J. Seray, 2009), 7–8.

6. Seray, *Richard Lesclide*, 97.

7. Seray, *Richard Lesclide*, 227–31.

8. Lesclide's office and the gymnasium of Eugène Paz (author of the chapter on exercise in the *Manual*) were located on the same street, rue des Martyrs, in Paris. On the exact publication dates of the *Almanac* and the *Manual* see Kobayashi, *Histoire du vélocipède*, 256.

9. Le Grand Jacques, *Manuel du vélocipède*, 6.

10. Seray, *Richard Lesclide*, 137–38.

11. And, more specifically to the French context, Lesclide's travel novel foreshadows *Le Tour de la France par deux enfants* (The Tour of France by Two Children), Augustine Fouillée's 1877 schoolbook, novel, and geography manual that was widely used in French schools well into the twentieth century, creating a sense of Frenchness for generations of schoolchildren.

12. Le Grand Jacques, *Le Tour du monde en vélocipède* (Paris: Aux Bureaux de la Publication, 1870), 271.

13. John Grand-Cateret writes that Benassit was born in London to French parents in 1834 (see *Les Mœurs et la caricature en France* [Paris: Librairie illustrée, 1888], 623), but the Bibliothèque nationale de France indicates that he was born in Bordeaux, France, in 1833.

14. For a summary of Benassit's work, see Grand-Cateret, *Les Mœurs et la caricature en France*, 623.

15. Only a handful of pre-1870 oil paintings of the velocipede exist. They are examined in detail by Scotford Lawrence in "Paintings of the Velocipede," in Besse and Henry, *Le Vélocipède: Objet de modernité*.

16. Emile Benassit, *Dragons à cheval*, Wikimedia Commons, https://commons.m.wikimedia.org/wiki/File:Louis_Emile_Benassit_(1833-1902)--Dragons_à_cheval.jpg.

17. George Sand and Rosa Bonheur, among others, sought authorization to wear pants in public in nineteenth-century France. See Patricia Marks, *Bicycles, Bangs, and Bloomers: The New Woman in the Popular Press* (Lexington: University Press of Kentucky, 1990). See also Rachel Mesch, *Before Trans: Three Gender Stories from Nineteenth-Century France* (Stanford: Stanford University Press, 2020). In 1896, doctor Léon Petit argued that the corset and the bicycle were strictly incompatible since the corset hampers breathing and hinders the intestines. He concluded, "Women desirous to take on the challenge of cycling would do well to get rid of an instrument [the corset] that . . . clashes with the masculine clothing they have so brazenly adopted" (Léon Petit, "La Bicyclette et le corset," *Paris-vélo almanach* [Paris: G. Charpentier and E. Fasquelle, 1896], 61).

18. Mikhail Bakhtin, *Rabelais and His World*, trans. Hélène Iswolsky (Bloomington: Indiana University Press, 1984).

19. Edward Muir, *Ritual in Early Modern Europe* (Cambridge: Cambridge University Press, 1997), 91.

20. The Bibliothèque nationale de France's online database, Gallica, indicates that this image dates from 1880. However, *La Chatte blanche* was a fairytale play written by the Cogniard brothers and first performed in 1852, then reprised at the Théâtre de la Gaîté in 1869. This poster announces that velocipedes from the Michaux Company were used during the play. Since the Michaux Company closed its doors in 1870, we are convinced that this advertisement was for the 1869 production, at the height of velocipedomania. https://gallica.bnf.fr/ark:/12148/btv1b90118598.r=velocipede?rk=85837;2.

21. Le Grand Jacques, *Manuel du vélocipède*, 77.

22. Le Grand Jacques, *Manuel du vélocipède*, 7.

23. Le Grand Jacques, *Manuel du vélocipède*, 6.

24. Le Grand Jacques, *Manuel du vélocipède*, 5.

25. Le Grand Jacques, *Manuel du vélocipède*, 39.

26. Le Grand Jacques, *Manuel du vélocipède*, 47.

27. Marc Olivier, "Civilization Inoculated: Nostalgia and the Marketing of Emerging Technologies," *Journal of Popular Culture* 44, no. 1 (2011).

28. See Corry Cropper, *Playing at Monarchy: Sport as Metaphor in Nineteenth-Century France* (Lincoln: University of Nebraska Press, 2008), chap. 6.

29. Le Grand Jacques, *Manuel du vélocipède*, 63.

30. Le Grand Jacques, *Manuel du vélocipède*, 5–6.

31. The train in Émile Zola's novel *La Bête humaine* offers a useful counterpoint here. The relationship between man and machine would be developed further with the advent of the automobile, again with resonance throughout French culture; for one such example see Octave Mirbeau's 1907 novel *La 628-E8*, in which the car takes on a life of its own. On this topic, see Claire Nettleton, "Driving Us Crazy: Fast Cars, Madness, and the Avant-Garde in Octave Mirbeau's *La 628-E8*," *Nineteenth-Century French Studies* 42, nos. 3–4 (2014). See also Corry Cropper's analysis of Alfred Jarry's 1902 novel *Le Surmâle*: "Like a Furnace: Alfred Jarry's *The Supermale*, Doping, and the Limits of Positivism," in *Culture on Two Wheels: The Bicycle in Literature and Film*, ed. Jeremy Withers and Daniel P. Shea (Lincoln: University of Nebraska Press, 2016).

32. Le Grand Jacques, *Manuel du vélocipède*, 34.

33. Le Grand Jacques, *Manuel du vélocipède*, 36. In his timidly erotic *Contes extragalants* (Paris: E. Dentu, 1886), Lesclide uses a similar metaphor, describing an actress's breasts as "snow-covered hills" (p. 56).

34. Le Grand Jacques, *Manuel du vélocipède*, 26.

35. Le Grand Jacques, *Manuel du vélocipède*, 40.

36. Le Grand Jacques, *Manuel du vélocipède*, 82.

37. "Si tu le veux, femme à l'œil fauve, / Je serai ton fauve lion, / Et je te ferai dans l'alcôve / La di-li-gen-ce de Lyon. / Trou-la-la ou . . ." See Paul Verlaine, *Œuvres poétiques complètes* (Paris: Robert Laffont, 1992), 627.

38. Richard Lesclide, *La Diligence de Lyon* (Brussels: Henry Kistemaeckers, 1882), 91.

39. Alfred Delvau, *Dictionnaire érotique moderne par un professeur de langue verte* (Basel: Karl Schmidt, 1864), 137–38.

40. For a sense of the great lengths to which Second Empire authors had to go to avoid censorship, see William Olmsted, *The Censorship Effect: Baudelaire, Flaubert, and the Formation of French Modernism* (Oxford: Oxford University Press, 2016).

41. Alfred Delvau, *Les Heures parisiennes* (Paris: Librairie Centrale, 1866), unnumbered page. In a second edition published in 1882 (Paris: C. Marpon et E. Flammarion), a cherub was added to cover the opening in the canopy and make the sketch appear a bit more innocent.

42. Laurence Klejman and Florence Rochefort, *L'Egalité en marche: Le Féminisme sous la troisième république* (Paris: des femmes, 1989), 48.

43. The Bois de Boulogne, home of the Pré Catelan, encouraged the coexistence of multiple discourses: open spaces in the modern city were the places of leisure and thus natural homes for the velocipede, and, at the same time, their wide expanses and their promised escape from the pressures of the city made them natural habitats for a variety of other leisure activities, including gendered and eroticized ones.

44. Marks, *Bicycles, Bangs, and Bloomers*, 184.

45. As mentioned in note 2, in our translation we have followed the original's capitalization of the word *velocipede* and its variants.

46. Throughout the text, the railway is consistently derided and even mocked, with the velocipede held up as a superior form of transportation.

47. "Le style est l'homme même." Georges-Louis de Buffon, *Discours sur le style et autres discours académiques* (Paris: Hachette, 1843), 11.

48. Maria Malibran (1808–36), a Spanish opera star, was born in Paris in 1808 and died in the UK in 1836 from injuries sustained when she fell from a horse.

49. The *Almanach des Vélocipèdes*, illustrated by "a jobless horse," was published the same year as the *Manual of the Velocipede* and includes earlier versions of some of this work's chapters.

50. The notion that art should instruct and delight was formulated in Horace's *Ars Poetica* (ca. 19 B.C., first translated into French by Jacques Peletier du Mans in 1541 and reprinted in 1545), particularly in the famous line that refers to literature as "miscuit utile dulci" (a mixture of useful and sweet).

51. The eighteenth-century naturalist George-Louis de Buffon wrote that the horse was man's noblest conquest. *Histoire naturelle, générale et particuliére*, vol. 4 (Paris: De l'imprimerie royale, 1753), 174.

52. Thinking about how to talk about the velocipede was a recurring theme in the French press; an article in *Le Journal amusant* in 1868 (see the introduction) introduced conjugations of the newly imagined verb *vélocipéder*, concluding, "What I love the most about inventions of modern engineering is their immediate consequences on the French language. They all enrich the dictionary with some barbarous verb, with some cacophonous noun, with some adjective that is impossible to pronounce." Arnould, "Vélocipédons," 2.

53. Writer, art critic, and urbanist Louis Petit de Bachaumont (1690–1771) is best known for his alleged role in the first volumes of the famous *Mémoires secrets pour servir à l'histoire de la République des Lettres* (18 vols., London: John Adamson, 1762–87), which describes literary life with considerable insider knowledge.

54. "Rider": the French word *cavalier* is commonly used, here and elsewhere, to designate the velocipede rider. It offers an echo of its equestrian roots and, like the term *gouvernail* (see note 57), is an example of grafting the new advancement of the velocipede onto more established practices, in order to soften its novelty and make it seem more familiar.

55. On the celerifere, see chapter 1, note 8.

56. Lexicographer and grammarian Louis-Nicolas Bescherelle (1802–83) is still well known in France for his important publications in the 1840s and 1850s. While today the name Bescherelle refers to the standard guide to twelve thousand French verb conjugations, in the nineteenth century it referred more commonly to his dictionary, the *Dictionnaire universel de la langue française.*

57. "Handlebar": the French word used throughout this chapter is *gouvernail.* Meaning "rudder," it highlights the extent to which the velocipede borrowed from other lexical fields—in this case, nautical—as it forged a new path into and through language.

58. *Dagobert and His Velocipede* is one example of a velocipede in the theater. See the introduction and the start of chapter 2 for a brief discussion of other examples.

59. "Fama": the god of fame and renown, known in Greek as Pheme or Ossa. Fama was the Latin equivalent. This is referring to riders standing on their saddles, while Amazons rode sidesaddle.

60. As a reminder, the rue des Martyrs that runs between Pigalle and the church Notre-Dame de Lorette is where the Paz gymnasium and Lesclide's office could be found.

61. As a result of the so-called Union latine (1865), the amount of silver in newly minted coins was reduced from 90 percent to 83.5 percent; consequently, people were keen to get rid of what they viewed as a coin worth less than its gold counterpart by passing it off to the state. An investigation begun in July 1868—announced by a formal decree (posted in the shop in this story)—revealed that the practice of paying taxes with five-franc silver coins was widespread. A government commission set out to further study the use of coins and whether to use only gold. In the present story, the shopkeeper hopes to show the relevant authorities that he prefers gold imperial coins, bearing Napoleon III's image, to silver or foreign ones. While many businesses routinely accepted silver and foreign coins, our Bonapartist merchant normally refuses them but is willing to take them from the protagonist because the velocipede is a big sale, and he can always use the coins to pay his taxes. For more on the real and perceived value of French coins during this time period, see Henry Willis, *A History of the Latin Monetary Union: A Study of International Monetary Action* (Chicago: University of Chicago Press, 1901; reprint, New York: Greenwood Press, 1968), especially 98–106.

62. An old type of firework today referred to as a black snake; after being lit, it begins to smoke and emit ash resembling a snake.

63. She is refusing to take coins from Italy and insists that payment be made with French coins bearing the image of the emperor, Napoleon III.

64. In his 1864 *Dictionnaire érotique moderne*, Alfred Delvau offers the definition of the lantern as a vagina, into which a man puts his "candle." He also defines "balloons" as "women's enormous buttocks, either natural or artificial, as many Parisian women had in the day, thanks to the crinoline."

65. "Game of graces": an early nineteenth-century game in which participants used dowels to catch and throw hoops in the air, called "jeu de grâces" in French; the echoes of gameplay and good graces are not lost in this tale of a courtship.

66. In Greek mythology, Polymnia (also spelled Polyhymnia) was the Muse of sacred poetry, hymn, dance, and eloquence.

67. In addition to the famous neighborhood in northern Paris, the other references in this sentence are to a small suburb fifteen kilometers from the center of Paris and to novelist Paul de Kock (1793–1871).

68. The Société des gens de lettres is an association, founded in 1838 by French authors Honoré de Balzac, Victor Hugo, Alexandre Dumas, and George Sand, that to this day promotes and protects authors' moral, legal, and financial rights.

69. First created in 1806 and celebrated during France's First and Second Empires, August 15 was Saint Napoleon Day. It honored a (previously inexistant) Saint Napoleon whose annual day was chosen to coincide with the birthday of Napoleon Bonaparte (August 15, 1769). During the Second Empire, the national holiday included a solemn mass at Notre Dame cathedral followed by free shows presented by imperial theater troupes, a national garden party, and light shows on major public buildings. In addition, the government would use the day as an opportunity to start a new civic calendar, including a number of new initiatives and gestures of goodwill such as donations to war orphans, widows, or the poor. The civic equivalent of the start of a new fiscal year, it was also the date when laws would be posted, go into effect, or expire. See, for example, *Programme de la fête nationale du 15 août 1855* (Paris: Gaittet et Cie).

70. Roughly fifty kilometers to the west of Paris, Mantes-la-Jolie is indeed a pleasant city, but it is the birthplace of neither Nantes-born writer Charles Monselet (1825–88) nor Italian-born (Savona) theater director Anténor Joly (1799–1852). Nevertheless, they were both fixtures on the rue des Martyrs, as attested in Charles Monselet's *Figures parisiennes* (Paris: Jules Dagneau, 1854): "Anténor Joly lived at 47, rue des Martyrs, in a vast house that resembled a phalansterian city. I went to see him there at least once a week" (95). Monselet also frequented the famous brasserie Le Divan japonais, located at no. 75, with friends Charles Baudelaire and Jules Vallès, among others.

71. "Prancing around and performing some fancy dressage moves": The original phrase *faire de la haute école* is used in equitation (as is its counterpart, *basse école*). *Haute école* jumps are advancing movements in which the horse leaves the ground; for this reason they are also referred to as "airs above the ground." Bernard Chiris, "Haute-Ecole et Basse-Ecole," 2009, http://www.cheval-haute-ecole.com/index16-04.html.

72. "Climbed up on my hobbyhorse": the French phrase *enfourcher son dada* also refers to returning to one's favorite topic of conversation.

73. In Ovid's *Metamorphoses*, the sea nymph Galatea fled from the cyclops Polyphemus who had just killed her lover Acis.

74. There is a route de la Vierge-aux-Berceaux (Virgin of the Cradles) in the Bois de Boulogne: that the road's name literally involves robbing cradles is another detail not lost in this love story.

75. The famous William Tell Overture comes from the opera *Guillaume Tell* (written by Gioachino Rossini to a libretto by Victor-Joseph Étienne de Jouy and L. F. Bis), first performed at the Salle Le Peletier in Paris in 1829.

76. A *Ranz des vaches* is a simple melody that was traditionally played on the horn by Swiss herdsmen as they led their cattle. One of the most famous *Ranz des vaches* appears in the overture to Rossini's *Guillaume Tell.*

77. Notre-Dame de Lorette is a well-known church at the foot of the rue des Martyrs. The Buttes-Chaumont is a public park, crisscrossed with paths and trails, that opened in 1867.

78. For commentary on this chapter, see Smethurst, *The Bicycle*, 55.

79. The Old French word *baston* (predecessor of *bâton*) appeared in the proverb "Autant vaut aller à pied que de chevaucher un baston maigre" (It is better to go on foot than to straddle a weak stick), a more modern version of which is "Mieux vaut aller à pied que de chevaucher un mauvais cheval" (It is better to go on foot than astride a bad horse). For more on the baton–horse lineage, see the website of the Centre de Recherche sur la Canne et le Bâton, http://www.crcb.org.

80. The author appears to be citing a retelling of the battle of Troy. If the original exists, we were unable to identify it.

81. "Aux temps heureux de la chevalerie" is a common folk song, with a tune on which many other songs are based (e.g., "sung to the tune of 'Aux temps heureux de la chevalerie'").

82. Denis Dominique Cardonne (1721–83) was Chair of Turkish and Persian at the Collège de France and author of a number of translations and historical works. François Pétis de la Croix (1653–1713) published the five volumes of his *Mille et un jours* (Thousand and One Days) between 1710 and 1712. Antoine Galland (1646–1715) famously translated (and added stories to) the *Thousand and One Nights*, as well as to *Aladdin or the Magic Lamp* and *Ali Baba and the Forty Thieves.*

83. Astolfo is a fictional character in a number of medieval tales of chivalry, notably in *La Chanson des quatre fils Aymon* (The Song of Aymon's Four Sons). He was later popularized as a humorous character in romance epics of the Italian Renaissance, in particular *Orlando Furioso* by Ludovico Ariosto (1474–1533), in which he is a carried off to the moon on a flying horse, or hippogriff.

84. This quotation comes from *Don Quixote* by Miguel de Cervantes. See this English edition for example: Miguel de Cervantes Saavedra, *The History of Don Quixote de la Mancha* (Hartford: Silas Andrus and Son, 1851), 310–11.

85. Soon after the scene from *Don Quixote* in the previous quotation comes this one, in which Don Quixote and Sancho Panza ride off high into the air, leading Quixote to tell his companion, "'Banish fear, my friend, the business

goes on swimmingly, with a gale fresh and fair behind us.' 'I think so too,' quoth Sancho; 'for I feel the wind here as if a thousand pairs of bellows were puffing at my tail.' And, indeed, this was the fact, as sundry large bellows were just then pouring upon them an artificial storm. . . . 'If we go on mounting at this rate, we shall soon be in the region of fire; and how to manage this peg I know not, so as to avoid mounting where we shall be burnt alive.'" (Cervantes, *History of Don Quixote*, 316).

86. This sentence is quoted almost verbatim from Viardot's note. Miguel de Cervantes Saavedra, *L'Ingénieux Hidalgo Don Quichotte de la Manche*, trans. and notes by Louis Viardot, vol. 2 (Paris: J.-J. Dubochet, 1836), 415n.

87. The archaic expression *à la mistanflûte* was used to describe something done extravagantly. The source is titled *Les Jeux de tous les âges au château de Robert mon oncle* (Paris: Librairie du Petit Journal, 1867). An anonymous publication from the same press as the *Manual of the Velocipede*, we suspect that it may have been authored by Le Grand Jacques himself, who could have used the quotation here as a way to promote his other work.

88. This essay was reprinted in *Le Vélocipède illustré*, April 1, 1869, 4. The Pré Catelan is a park in Paris's Bois de Boulogne. After being quarried from 1853 to 1858 in order to pave the paths through the Bois, the area was turned into a leisure park, complete with concert hall, brasserie, aquarium, exhibit space, and the 1,800-seat Flower Theatre, all surrounded with landscaped gardens designed by Gabriel Davioud and Jean-Pierre Barillet-Deschamps. While its regular cultural activities ceased with the onset of the Franco-Prussian War in 1870 and the Paris Commune in 1871, the Pré Catelan was the location of the height of Parisian leisure during the Second Empire, its long pathways ideal for strolling and for riding velocipedes. Théobald de Saint-Félix was the director of the Pré Catelan from 1866 until the fall of 1869. When his tenure came to an end, he published a pamphlet denouncing the administration for their lack of thought about the Pré Catelan and for their elitist and petty policy decisions. His text, *La Comédie des hommes et des chiffres au bois de Boulogne* (Paris: 1870), bears the epigraph, "The justice of the people is slow but certain." It also indicates the numerous challenges he faced from other city officials as he tried to encourage the creation of a velocipede community.

89. The first velocipede race was held on December 8, 1867. It started in Paris at the Panorama National (today the site of the Théâtre du Rond-Point) near the Michaux workshop (29, avenue Montaigne), crossed the Seine from the Champ de Mars, where the second Exposition universelle had just ended (April 1–November 3, 1867), and finished at the place d'Armes, in front of the main entrance to the Château de Versailles. As reported in the *Courrier de la Drôme et de l'Ardèche* (Abel Céas, December 18, 1867), between 120 and 150 riders left at ten o'clock. The winner covered the seventeen kilometers in just under an hour. After some races in southern France in 1868 (Cannes, March 18; Hyères, April 13; La Réole, May 21), the second race to take place in Paris was announced for the Pré Catelan on May 24, although it is not clear that it ever took place. It was supposed to be presided over by Eugène Paz, who in 1859 had founded a gymnastics club in Paris, Les Amis de la gymnastique. The Pré Catelan hosted nearly a dozen

races in 1868–69, whether in May or not, one of which was the subject of Jules Peloq's drawing in *L'Univers illustré* on June 6, 1868, "Une course de vélocipèdes" (p. 348).

For some, the first significant velocipede race in Paris took place on May 31, 1868, in the parc de Saint-Cloud, on the grounds of the château, in the presence of Louis-Napoleon, the son of Emperor Napoleon III (see the introduction). A medal bearing the effigy of the emperor was even given to the winner, Englishman James Moore (Kobayashi, *Histoire du vélocipède*, 241). The Touring club de France commemorated Moore's victory with a plaque seventy years later in 1938. While plans for a centenary celebration were canceled because of the political and social unrest of May 1968, for the sesquicentenary a handful of riders from around the world recreated the race, on velocipedes and in period dress. For a description of the first velocipede races, see Herlihy, *Bicycle: The History*, 96; Andrew Ritchie, *Early Bicycles and the Quest for Speed*, 2nd ed. (Jefferson, NC: McFarland, 2018), 326; Kobayashi, *Histoire du vélocipède*, 239–40; and Francis Robin, *Le Paris-Versailles du dimanche 8 décembre 1867: Première Course cycliste de tous les temps?* (Pomeys: Vélocithèque, 2017). Accounts of early velocipede races are abundant in the French press of the time; see, for example, *Le Petit Journal* (June 2, 1868); *L'Univers illustré* (June 6, 1868); and *Le Vélocipède illustré* (May 24, 1870).

90. The first race of *vélocipédiennes*, held in Bordeaux in November 1868 and won by Mademoiselle Julie in front of over three thousand spectators, was the subject of an illustration in *Le Monde illustré* on November 21, 1868. See the image (figure I.6) and a discussion of the race in the introduction.

91. The first French veloce club was founded in Valence in March 1868; its Parisian counterpart followed in May. See Kobayashi, *Histoire du vélocipède*, especially part 4, "Les Courses de Vélocipèdes, 1867–1870"; and Ritchie, *Early Bicycles*, 326n14. See also Alex Poyer, *Les Premiers Temps des véloce-clubs* (Paris: L'Harmattan, 2003).

92. According to Greek mythology (and as retold to French readers by François Fénelon in the seventeenth century), Idomeneus led the Cretan armies during the Trojan War. Jean Joseph Jacotot (1770–1840) developed a pedagogical method of intellectual emancipation, according to which pupils differ only according to their will to use their intelligence. To a student learning a new language, he would give a short passage and encourage the pupil to study all aspects of it—words, syntax, meaning—and to draw as much as possible about the entire language from that initial paragraph. More recently, Jacotot's method and principles are discussed in Jacques Rancière, *The Ignorant Schoolmaster: Five Lessons in Intellectual Emancipation* (Palo Alto: Stanford University Press, 1991).

93. Over multiple regimes throughout the nineteenth century, the financing of railways was plagued with speculation and scandal, often in the form of bonds that were then invested, unpaid dividends, or stocks sold that had no real value. A report commissioned by the Ministry of Public Works and authored by Alfred Picard (who had been the commissioner of the 1900 Olympics) looks back at the entire period and outlines a number of the abuses. Alfred Picard, *Les Chemins de fer: Aperçu historique, résultats généraux de l'ouverture des chemins de fer,*

concurrence des voies ferrées entre elles et avec la navigation (Paris: H. Dunod et E. Pinat, 1918), particularly pp. 411–12. Le Grand Jacques criticized the railways in his newspaper, *La Vitesse*, accusing them of operating as "a monopoly, becoming a plague in our era" (July 23, 1871, 3).

94. Mouchette was a common name for characters in vaudevilles in the nineteenth century. See, for example, Antoine Simmonin and Théodore Nézel's *L'Ane mort et la femme guillotinée* (1832), Eugène Labiche's *Les Noces de Bouchencoeur* (1857), and Louis Leroy and Alfred Delacour's *Les Mormons à Paris* (1874). The names in this chapter immediately plunge the reader into the world of comic theater and farce.

95. Napoleon I suffered from recurrent abdominal distress, and it is believed that he died of gastric cancer. See Alessandro Lugli et al., "Napoleon Bonaparte's Gastric Cancer: A Clinicopathologic Approach to Staging, Pathogenesis, and Etiology," *Nature Clinical Practice Gastroenterology and Hepatology* 4 (January 2007): 52–57.

96. Pierrot is a conventional clown-like stock character associated with Carnival. Lesclide himself used Pierrot in two plays he authored.

97. In the prologue of his *Tiers Livre*, François Rabelais referred to the Germans and Swiss as *Lifrelofres*. Notes in editions as early as 1711 include the explanation that Rabelais used this expression because when they spoke it sounded to him as though they only ever said "liffre loffre." See François Rabelais, *Œuvres de maître François Rabelais publié sous le titre de Faits et dits du géant Gargantua et de son fils Pantagruel* (Amsterdam: Henri Bordesius, 1711), 44.

98. This is a reference to the famous cartoon that appeared in the satirical weekly *La Caricature morale, politique et littéraire*, depicting King Louis Philippe's face morphing into a pear. From an initial sketch by *La Caricature*'s director Charles Philipon in 1831, Honoré Daumier's rendering helped solidify the image that would be a recurring theme throughout the July Monarchy (1830–48). As William Makepeace Thackeray recalled in *The Paris Sketch Book*, vol. 2 (New York: D. Appleton, 1852), 23, "Everyone who was at Paris a few years since must recollect the famous '*poire*' which was chalked upon all the walls of the city and which bore so ludicrous a resemblance to Louis Philippe. . . . *La Poire* is immortal."

99. Licorice root was used in many popular healing remedies in France. See Le Sage, *Recueil des plus beaux secrets des grands guérisseurs* (Paris: Paul Leymaire, 1931), 32.

100. In myth, Lemnos is the island of Hephaestus (the Greek equivalent of the Roman Vulcan), the god of metallurgy and blacksmith to the gods. The Parisian equivalent is obviously Michaux.

101. While not explicitly termed a *fuite*, this scene offers at least an implicit reference to the *fuite à Varennes* (flight to Varennes), when King Louis XVI and Marie Antoinette unsuccessfully attempted to flee from Paris during the night of June 20–21, 1791. Their hopes of arriving in Montmédy to lead forces ready for a counterrevolution were dashed when they were captured in Varennes. Their failed attempt significantly increased public hostility toward the monarchy, as well as to the monarch more specifically, leading to his execution in 1793.

102. In Voltaire's satirical novel *Candide* (1759), philosopher Pangloss repeats his famous refrain, "Everything is for the best in the best possible world."

103. The phrase *va bon train* could be translated as "goes like blazes." Pseudonym of journalist Philippe Dubois (1862–1918), Dr. Vabontrain was the author of *Manuel de santé tintamarresque du docteur Vabontrain, ou Cours d'idiopathie à l'usage des gens du monde* (Paris and Tours: impr. de E. Mazereau, l'an des vélocipèdes, 1869).

104. This chapter's title is "Causerie sur la gymnastique," literally a chat or discussion on the topic of gymnastics. In France in the 1860s, *gymnastique* was a generic term that meant exercise for fitness, strength conditioning, and agility. Large gymnasiums taught weight training, fencing, boxing, horseback riding, and eventually velocipede riding. After the Franco-Prussian War of 1870, the term *gymnastique* became synonymous with military exercises and discipline.

105. This passage on the shower and hydrotherapy would be reprinted, nearly word for word, by Eugène Paz in an article in the weekly magazine *La Bicyclette*, "De l'hydrothérapie," July 7, 1893, 1352–53.

106. The "keyboard of the human mechanism" reminds readers of de La Mettrie's 1748 text *L'Homme machine* (Man a Machine), in which he compares the human machine to a harpsicord whose chords can be plucked to give voice to the entire human machine. Julien Offray de La Mettrie, *L'Homme machine* (Paris: Elie Luzac fils, 1748), 34.

107. Champion of *la gymnastique*, Eugène Paz (1837–1901) was the author of a number of studies, including *La Santé de l'esprit et du corps par la gymnastique: Etude sur les exercices du corps depuis les temps les plus reculés jusqu'à nos jours, leurs progrès, leurs effets merveilleux, leurs diverses applications et leur combinaison avec l'hydrotérapie* [*sic*], published in Paris in 1865 by the Librairie du Petit Journal (Paz was also one of the directors of *Le Petit Journal*). Émile Zola reviewed Paz's publication in "La Littérature et la gymnastique," published in the "Revue littéraire" section of *Salut public* (published in Lyon, October 5, 1865), later collected in his *Mes haines*. Ten days after Zola's article was published, Paz opened his large gymnasium at 40, rue des Martyrs in Paris's ninth arrondissement. He adopted velocipedes as part of his gymnastic training; see the *Note on Monsieur Michaux's Velocipede*, section 2, "Practical Applications." In 1873 Paz would also establish the Union des sociétés de gymnastique de France.

108. The original refers to the women as *écuyères*, that is, equestrians or jockeys.

109. Isabelle is referencing the fact that after the French defeat at Waterloo, the British kept Napoleon prisoner on Saint Helena from 1815 until his death in 1821.

110. The French reads "des courses de fatigue," literally "fatigue races." This is reminiscent (again) of horses that were grouped into racing horses, hunting horses, and workhorses, called *chevaux de fatigue*.

111. Bernard asks how much for the "bouquetière," a double entendre between the vase and the woman selling the flowers.

112. "Le Beau Dunois" is another name for the song titled "Partant pour la Syrie" by Hortense de Beauharnais (text by Alexandre Laborde), the mother of

Emperor Napoleon III. The song became the imperial anthem during the Second Empire.

113. The French *rameneur* is slang for a man who tries to hide his baldness, combing his hair over his balding spot. It can also refer to a shill, but that is an unlikely translation of the word in this context.

114. This is a reference to the fact that the Rhineland, from the time of the fall of Napoleon I and the implementation of the Congress of Vienna, was controlled by Prussia. For details about the buildup to the 1870 Franco-Prussian War, see David Wetzel, *A Duel of Giants: Bismarck, Napoleon III, and the Origins of the Franco-Prussian War* (Madison: University of Wisconsin Press, 2003). Hippolyte simply means that Félicie is on a rival team.

115. Rosette and Hortensia are on different teams.

116. This is a reference to Virgil's *Aeneid* (bk. 4), in which Hyrcania (modern-day Iran) is referred to as a land filled with wild tigers. The "tigress of Hyrcania" was famously put in verse by Paul Verlaine in his poem "Dans la grotte," published the same year as the *Manual of the Velocipede* (1869) in *Fêtes galantes* ("Et la tigresse épouvantable d'Hyrcanie").

117. A louis is a gold coin minted under Napoleon. Twenty-five louis represents a very hefty sum.

118. Founded in November 1868, La Société des vélocipèdes was the fourth-oldest cycling club in France; see Dauncey, *French Cycling*, 17.

119. Richard Lovelace (1617–57) was a poet who fought in support of King Charles I during the English Civil War. His best-known lines come from the end of his poem "To Lucasta, Going to the Warres": "I could not love thee, dear, so much, / Lov'd I not Honour more." See *Lucasta* (London: John Russell Smith, 1864), 27.

120. According to legend, Archimedes worked out how to measure volumes of irregular objects by submerging himself naked in his bath. He was so thrilled at his discovery that he ran through the streets shouting, "Eureka!"

121. We have translated the word *bicycle* as "two-wheeler": the term *bicycle* was used to differentiate a two-wheeled velocipede from a three- or four-wheeled one.

122. Bismarck brown, named for the German statesman Prince Otto von Bismarck (1815–98), was a color that enjoyed tremendous popularity in the mid-1860s before it became a standard color in dyes (called *vesuvin* when used as a biological stain). Mathilde Mauté mentioned it in her diary: "The year 1867 was another good year for me. Mother often took me to the Universal Exhibition. I remember having met the Empress there a few times. It's thanks to her incredible beauty that she was able to pull off the horrible fashion of that year: disgraceful crinolines; unrefined, fake hair extensions; ridiculous little hats; long ribbons called 'follow me, good looking!'; short, thick waists; boots with pompoms. It wasn't limited to the colors, so garish it makes you want to scream, ferocious blue, raw green, blood red, and a yellowish brown called 'Bismarck.'" Mathilde Mauté [L'Ex-Madame Paul Verlaine, *Mémoires de ma vie*, ed. Michael Pakenham (Seyssel: Champ Vallon, 1992), 58–59.

123. The Café Anglais was a famous restaurant located at 13, boulevard des Italiens. Chef Adolphe Dugléré (1805–84) built the café's reputation for fine

dining, and on June 7, 1867, he hosted a banquet for Tsar Alexander II of Russia and Kaiser Wilhelm I and Otto von Bismarck of Prussia, all of whom were visiting Paris for the Exposition universelle. Now referred to as the "Three Emperors' Dinner," the meal consisted of sixteen courses, with eight wines served over eight hours. The Café Anglais is commonly referenced in literature of the period, including in Honoré de Balzac's *Old Goriot*, Gustave Flaubert's *Sentimental Education*, Émile Zola's *Nana*, Guy de Maupassant's "The False Gems," and Marcel Proust's *In Search of Lost Time*, among many others.

124. This is a not-so-subtle reference to the red flag used by socialists during the 1848 Revolution.

125. Marie-Anne Lenormand (1772–1843) was a famous fortune-teller during the Napoleonic era; the figures referred to here would have appeared on the cards that she drew to tell fortunes. Sébastien Érard (1752–1831) was a Parisian manufacturer of pianos and harps.

126. In Italian in the original; in English, it means "Who knows?"

127. He appears to be working out a rudimentary sketch problem. "Problems" like this were published regularly in Le Grand Jacques's newspaper, *Le Vélocipède illustré*. See two examples on p. 3 of the issue of October 3, 1869. The editor promised to publish the names of any readers who sent in the correct solution.

128. While "orfraie" means white-tailed eagle, the bird of prey is central to the expression for screaming bloody murder, *pousser des cris d'orfraie*. We have kept the count's alliterative name in French. While not necessarily relevant to this story, references to the war being waged in South America between Paraguay and the Triple Alliance of Argentina, Brazil, and Uruguay would doubtless have been familiar to readers in 1869; the war had begun in 1864 and would end in early 1870.

129. The French original retains the Latin *aurea mediocritas*, referring to the golden mean, or the happy medium, appearing early on in the Delphic maxim "Nothing to excess" and later central to Aristotelian philosophy as the preferred middle between two extremes.

130. The French equivalent of "they lived happily ever after" is "they lived happily and had many children." This story is ironically playing with that formulaic conclusion.

131. Journalist Jules Rohaut wrote under the pseudonyms Jules Dementhe, Jean Lhuillier, and John Stick. In a letter dated February 20, 1872, Gustave Flaubert described him as "a very intelligent man, probing and able to do anything, from gossip columns to satire in verse." (See the letter to Charles-Edmond Chojecki housed online by the Université de Rouen and edited by Yvan Leclerc et Danielle Girard: https://flaubert-v1.univ-rouen.fr/correspondance/edition/index.php.)

132. George Brian Brummell (1778–1840), a well-known dandy, was born in England in 1778 and died in France in 1840. Novelist Honoré de Balzac called him a "prince of fashion" whose "philosophy" consisted of connecting the "elegant life" to "the perfection of human society" (*Traité de la vie élégante* [Paris: Librairie Nouvelle, 1854], 38). Author and critic Jules Barbey-d'Aurévilly devoted an entire book to Brummell. In it he claimed that "women couldn't forgive him

for having as much grace as them; men despised him because they couldn't be like him" (*Du dandysme et de G. Brummell* [Caen: B. Mancel, 1845], 48).

133. Abbé Dupanloup (1802–78) was a French ecclesiastic, elected to the Académie française in 1854 (he resigned in 1875 upon the election of Émile Littré, an agnostic). Louis Veuillot (1813–83) was a conservative journalist known for his stance in favor of the sovereignty of the pope, particularly as editor of *L'Univers*, which had just been reinstated in 1867 (Veuillot had been its editor from 1848 until it was suppressed in 1860). The battle between the two of them played out in the press in November 1869; see John O'Malley, *Vatican I: The Council and the Making of the Ultramontane Church* (Cambridge, MA: Harvard University Press, 2018). In short, the author is satirizing their prudish conservatism. The first issue of *Le Vélocipède illustré* in 1869 contains several references to Veuillot, who had just published his own satirical poems, *Les Couleuvres* (Paris: V. Palmé, 1869). In Veuillot's collection, a poem entitled "Un débutant" (A Beginner; 106) presents a poetic subject scolding a young boy whose poetry is not yet worthy. The versified response in *Le Vélocipède illustré* turns the tables on Veuillot, whose own verses are judged to be seriously lacking (R. T., "A Monsieur Louis Veuillot," April 1, 1869).

134. Lesclide reprinted a version of this story in his *Contes extragalants*, under the title "La Fin d'un comte" (The End of a Count), which begins, "Here is the true and disturbing story of Count Raoul de Rochefort, who lived in the year of our lord 1875, in a land rich with peanuts, who became a member of the Jockey Club and who was led on a path, full of detours but safe, to the abyss of marriage, from which God protects you as long as you are a bachelor" (107–08). In this 1886 rewriting of the story, Lesclide opted to replace velocipedes with horses. *Ecarté* was a nineteenth-century card game similar to whist or euchre.

135. The French expression *petits crevés*, translated here as "pretty boys," has echoes elsewhere in the cycling lexicon, as a number of words related to flat tires come from *crever*, meaning "to burst" or "to puncture."

136. A reference to Jules Verne's *Les Aventures du capitaine Hatteras*, which began appearing in serial form in the *Magasin d'éducation et de récréation* on March 20, 1864 (first part, *Les Anglais au pôle Nord*) and then again beginning on March 5, 1865 (*Le Désert de glace*). The complete version was first published on November 26, 1866, under the title *Voyages et aventures du capitaine Hatteras* (Paris: Hetzel).

137. "Peanuts": the French *nèfle*, also called *cul-de-chien* because of its appearance, literally means "medlar" (*Mespilus germanica*), a fruit-bearing tree in the rose family indigenous to southwest Asia and southeastern Europe. Partly because the fruit rots before it ripens, in literature it appears figuratively as a symbol of prostitution, in addition to being slang for female genitalia. Its plural form used here, *des nèfles*, refers figuratively to a pittance or peanuts.

138. Writer Arsène Houssaye (1815–96), to whom Baudelaire dedicated his collection of prose poems entitled *Le Spleen de Paris*, was the editor of a number of literary journals. The reference to hiding places refers to the moment in 1863 when, while excavating the site of the chapel of Saint-Florentin at the Château d'Amboise (Loire Valley), Houssaye found a partially complete skeleton and

stone fragments bearing the inscription "EO . . . DUS VINC." He concluded that they were the remains of Leonardo de Vinci, which had been reinterred in the chapel of Saint-Hubert at the same château. The unproven claim ("Biology in Art: Genetic Detectives ID Microbes Suspected of Slowly Ruining Humanity's Treasures," JCVI, June 18, 2020, https://www.jcvi.org/leonardo-da-vinci-dna-project) led Houssaye to write *Histoire de Léonard de Vinci*, published by Dentu in 1869, the same year the *Manual of the Velocipede* was published.

139. "Infantrymen": during the Battle of Trocadero, fought in Cádiz, Spain, on August 31, 1823, French forces invaded Spain, defeated Spanish liberal forces, and restored the absolute rule of King Ferdinand VII. The word *voltigeur* means both "circus performer/acrobat" and "infantryman." Given the reference to Trocadero, the military reference is clear, although the carnivalesque is never far behind. Finally, according to Delvau's *Dictionnaire érotique moderne*, *le petit voltigeur*, or "little infantryman" (who "enlists voluntarily"), refers to the phallus.

140. "Coat of arms": the word *blason*, for "family crest" or "coat-of-arms," also designates a kind of poem that dates back to the late fifteenth century (Guillaume Alexis, *Le Grant Blason des faulces amours*) and was popularized in the sixteenth century thanks to Clément Marot's *Beau Tétin* (1535). The *blason* is a short poem (thirty to forty lines) in octosyllabic or decasyllabic lines that celebrates a specific part of a woman's body.

141. *Blooméristes*, or women in trousers, were the subject of a well-known comic play, *Les Blooméristes ou la réforme des jupons*, that premiered in Paris in 1852. On the subject of bloomers in France, see Pascale Gorguet Ballesteros, "Women in Trousers: Henriette d'Angeville, a French Pioneer?" *Fashion Practice* 9, no. 2 (2017): 200–213.

142. Quotation from Pierre Corneille's play *Le Cid* (1637), act 2, scene 2, in which Don Rodrigue says that his fighting qualities and ambitions are such that he will be victorious with his first strike and has no need of practice: "Mes pareils à deux fois ne se font point connaître, / Et pour leurs coups d'essai veulent des coups de maître." Pipette changes the first pronoun from "my" to "your."

143. *Poupée*, "doll," according to Delvau's *Dictionnaire érotique moderne*, was slang for a promiscuous woman or a prostitute.

144. Singer, actor, and librettist Jean Elleviou (1769–1842) betrayed his family's desire for him to follow in his father's footsteps and become a doctor. When they learned of his intention to become an actor, they had him arrested and redirected to the medical profession. After Elleviou was sent back to Paris to resume his studies, he quickly abandoned them and began his career on the stage, debuting as a bass in Pierre-Alexandre Monsigny's *Le Déserteur* in the Opéra-Comique's Salle Favart. Elleviou is the subject of some apocryphal stories—including one in which his family had him held in a prison tower, from which he sang, from the romance *Richard Cœur-de-lion*, the lines "In a dark tower / a powerful king languishes" (*Dictionnaire de la conversation et de la lecture*, ed. M. W. Duckett, vol. 8 [Paris: M. Lévy Frères, s.v. "Elleviou")—but all accounts refer to his practically overnight transformation from bass to tenor, in which capacity he immediately distinguished himself in the role of Philippe in Nicolas Dalayrac's *Philippe et Georgette*.

145. "Curled up": the French verb is *pelotonner*, which can also mean forming into a compact group. This word would give rise to the cycling term *peloton*.

146. In a full-page parody of the podoscaphs published in the satirical newspaper *Paris-Caprice* in 1869 (figure 3.29), the text under the first image depicting a woman falling off a podoscaph reads, "The nautical velocipede is unsinkable, but only the velocipede [though the rider may fall into the water!]." The bottom left shows a man and woman together on a podoscaph. The dialogue below the illustration reads, "'But monsieur, I can't go back to the beach; I'm sure Mama saw us.' 'Well, mademoiselle, let's go to England.'" And the woman lower right explains, "Velocipedes on the water are sad; I think it's best to leave them to the provinces." An article in the daily paper, *La Petite Presse*, notes that races were held on the Saône River in Mâcon (a city just north of Lyon), with maritime velocipedes traveling both upstream and downstream (A. Merlé, "La Fête mâconnaise," *La Petite Presse*, May 19, 1869, 3).

147. The original proverb, attested by François Guizot (*Nouveau Dictionnaire universel des synonymes* [Paris: Aimé Payen, 1848], 443), is *Le gibet* (or in some cases *justice* or *dieu*) *ne perd jamais ses droits* (The scaffold/justice never loses its claim), meaning that criminals or sinners will always be caught or will suffer in some manner for their crimes. Here, the author substitutes "prose" for "scaffold," meaning that realism has its claim, that it is now time to address the cold facts of velocipede ownership.

148. The rue Jean Goujon runs from the place de la Reine Astrid to the place François Ier. Michaux's address is also sometimes given as the avenue Montaigne, a street that paralleled the rue Jean-Goujon. It appears that the storefront was on avenue Montaigne and the workshop one street away, on the rue Jean-Goujon. The Michaux Company would change its name to Compagnie Parisienne later in 1869.

149. Wheel hubs were equipped with small oil-filled cups or vials that would slowly release oil onto the axle, keeping the wheels well lubricated on long rides.

150. Le Grand Jacques's newspaper, *Le Vélocipède illustré*, carried advertisements for lights that could be opened or closed and removed and stored in saddle bags. See, for example, the last page of the issue dated September 4, 1870.

151. Odometers were used on public coaches as early as the seventeenth century and were adapted for the velocipede. See Edouard Fournier, *Le Vieux-Neuf: Histoire ancienne des inventions et découvertes modernes*, vol. 2 (Paris: E. Dentu, 1877), 375. For a description of velocipede odometers see Kobayashi, *Histoire du vélocipède*, 219–20. Odometers were advertised in Le Grand Jacques's newspaper, *Le Vélocipède illustré*; for example, the back page of the issue published on September 4, 1870, listed two: a *vélocimètre* (with two dials for meters and kilometers traveled) for twenty-five francs or a simpler odometer for sixteen francs.

152. In addition to the brand of a mid-twentieth-century bicycle company, the unattributed name NOLY is obviously an anagram of Lyon; as such, it seems to point to the Lyon-based Olivier brothers, whose financial support got Michaux velocipedes up and running in 1868–69. After a split from Pierre Michaux, the Olivier brothers continued without him and instituted mass manufacturing:

the dozen machines that sixty workers produced in a day became one hundred to two hundred machines made by three hundred workers, and the increased demand required additional factories outside of Paris. See Alexandre Sumpf, "La Naissance de l'industrie du vélo," L'Histoire par l'image, July 2011, http://www.histoire-image.org/fr/etudes/naissance-industrie-velo; and Henri Vigne, "Le Vélocipède," *L'Illustration: Journal universel* 53 (1869): 382.

CHAPTER FOUR VELOCIPEDOMANIA IN VERSE

1. Théodore de Banville, "L'Homme vélocipède," published in the satirical newspaper *Le Charivari*, July 24, 1868, later included as part of his "Triolets," in *Les Occidentales* (Paris: Alphonse Lemerre, 1875).

2. We have limited the poems in this brief anthology to poems about the velocipede. Beginning in 1890, there were a great number of poems about its successor, *la bicyclette*; they include Édouard de Perrodil, "La Bicyclette" (*Les Echos*, Paris: Vanier, 1891, 17); and verses that Richard Lesclide wrote in the beginning of his letter-preface to *Pédalons!* by Jehan de la Pédale, with a preface by Richard O'Monroy (Paris: Bureaux du "Véloce-Sport," 1892, 13). The most substantial is Jules Riol's thirty-two page *La Bicyclette, monologues en vers, dédié au Touring-Club de France* (Paris: Lanée, 1896), of which excerpts are included in Edward Nye's anthology *À bicyclette* (Paris: Les Belles Lettres, 2000; 2nd ed., 2013), 87–93.

3. This untitled poem is quoted in Baudry de Saunier, *Histoire générale de la vélocipédie*, 47–8; and in G. Meyland, "L'Histoire des sports," *Le Radical*, October 4, 1904, 7.

4. A brief life-sketch of Raoul Suérus can be found in "Actes de la société," *Bulletin de la Société de géographie de Lille*, no. 4 (1930): 193. It bears mentioning that the same night that Suérus read this poem in the Latin Quarter in Paris, the not-yet-famous Guy de Maupassant read a poem at his *lycée* in Rouen. Maupassant's "Saint-Charlemagne" can be found online here: http://maupassant.free.fr/poesie/charlemagne.html.

5. In Roman satire, "Trossulus" is a contemptible nickname applied to an Etruscan coxcomb or a dandy fop. See A. Riese's edition of Marcus Terrentius Varro's *Menippean Satire* (*M. Terentii Varronis Saturarum Menippearum Reliquiae* [Berlin: Teubner, 1865]), 480.1.

6. Euterpe is the muse of light music and lyric poetry, arts better suited to small venues. This is likely a reference to the construction of the new Théâtre du Vaudeville (now the Gaumont Opéra cinema) with its large classical façade that was completed in 1868.

7. In 1869, the feminist journal *Le Droit des femmes* published a manifesto signed by a number of women. "Its objective [was] to mobilize public opinion in favor of the 'civil rights of women,' access to a secondary and university education, the right to work and equality of salaries." See Laurence Klejman and Florence Rochefort, *L'Egalité en marche: Le Féminisme sous la troisième république* (Paris: Des Femmes, 1989), 48. This is yet another sign of societal change evoked by the velocipede.

8. Introduced in 1866, Chassepot rifles were the first bolt-action, breech-loading rifles used by the French military. They are evoked here as a symbol of technological progress.

9. "Pallas" indicates Minerva (Athena), the clever goddess of strategy, who, according to myth, was born in full armor directly from the head of her father Jupiter (Zeus).

10. The phrase in Latin would be translated more precisely as "the middle of a man sits upon the saddle."

11. Ixion suffers eternal torment among the dead (in Hades), bound to a continuously revolving wheel of fire as punishment for having attempted to rape Juno (Hera).

12. The purposefully stilted English here reflects the archaic quality of the epic word used in the source text, *sonipes*.

13. The Latin text almost certainly demands emendation—*cursui afferre modum*—for the sake of meter; however, the sense of the sentence remains obscure.

14. A reference to the holiday of Mardi Gras (Carnival) that would have been celebrated in Paris the following weekend.

15. "O'er them" is intentionally archaic to match the archaizing *ollis*.

16. This is a clever mythological trope. The winged horse Pegasus plays two roles in classical mythology, actively transporting Jupiter's (Zeus's) thunderbolts back to Olympus and occasionally offering poets and heroes passage to the gods' presence as well. By "Phoebus's wild peaks," the speaker means Mt. Helicon, the home of the Muses where Phoebus Apollo presides over them. Sucrus's conclusion claims that the velocipede has outstripped Pegasus as the best vehicle to immortality.

CONCLUSION

1. Kobayashi, *Histoire du vélocipède*, 311.

2. Le Grand Jacques, "Faits divers," *La Vitesse*, 16 July 1871, 3.

3. Charles D . . . "Journal d'un véloceman pendant le siège de Paris," *La Vitesse*, July 23, 1871.

4. Charles D . . . "Journal d'un véloceman," July 23, 1871.

5. Charles D . . . "Journal d'un véloceman pendant le siège de Paris," *La Vitesse*, July 30, 1871.

6. Le Grand Jacques, "Informations," *La Vitesse*, August 13, 1871.

7. Le Grand Jacques, "A nos amis," *Le Vélocipède illustré*, May 2, 1872.

8. Rémy Lamon, *Théorie vélocipédique et pratique* (Paris: Imprimerie Nouvelle, 1872), 7.

9. Lamon, *Théorie vélocipédique et pratique*, 11.

10. It is worth noting that while the "safety" bicycle also caught on in the United States in the 1890s, it was largely abandoned by the early twentieth century. By contrast, the bicycle became an essential part of daily life in France and in other European countries, most notably Belgium and Flanders. One could rightly argue that geography (the size of France and the relatively shorter distance between villages meant the bicycle was a more practical mode of transportation)

and economics (a robust American economy made the automobile a more attractive option) were both factors. Our contention is that velocipedomania also played a significant role: the passion for the two-wheeler was more deeply embedded in France and reflected French cultural values in a way that enabled it to become the embodiment of French identity.

11. She also posed on a bicycle (and wrote a poem) for an advertisement for the popular Coca Mariani, a French tonic wine that was advertised to cyclists as an "elixir" promising to confer strength and youth.

12. Jehan de la Pédale, *Pédalons*, 13.

BIBLIOGRAPHY

"Actes de la société." *Bulletin de la Société de géographie de Lille*, no. 4 (1930): 191–93.

Arnaud, Pierre. *Le Militaire, l'écolier, le gymnaste: Naissance de l'éducation physique en France (1869–1889)*. Lyon: Presses universitaires de Lyon, 1991.

Arnould, Arthur. "Vélocipédons." *Le Journal amusant*, June 13, 1868.

Aubray, Maxime. "Vélocipède & Hippophagie." *Le Petit Marseillais*, January 8, 1869.

Avenier, Cédric. "Alfred Berruyer (1819–1901): La Volonté d'un architecte diocésain." *La Pierre et l'écrit* 13 (2002): 69–95.

Bachaumont, Louis Petit de. *Mémoires secrets pour servir à l'histoire de la République des Lettres*. 18 vols. London: John Adamson, 1762–87.

Baker, Alan R. H. *Amateur Musical Societies and Sports Clubs in Provincial France, 1848–1914: Harmony and Hostility*. London: Palgrave Macmillan, 2017.

Bakhtin, Mikhail. *Rabelais and His World*. Translated by Hélène Iswolsky. Bloomington: Indiana University Press, 1984.

Ballesteros, Pascale Gorguet. "Women in Trousers: Henriette d'Angeville, a French Pioneer?" *Fashion Practice* 9, no. 2 (2017): 200–13.

Balzac, Honoré de. *Traité de la vie élégante*. Paris: Librairie Nouvelle, 1854.

Banville, Théodore de. "Le Vélocipède." In *Les Occidentales*. Paris: Alphonse Lemerre, 1875.

Barbey-d'Aurévilly, Jules. *Du dandysme et de G. Brummell*. Caen: B. Mancel, 1845.

———. *Le Théâtre contemporain*. Vol. 3. Paris: Maison Quantin, 1889.

Bard, Christine. *Une histoire politique du pantalon*. Paris: Editions du Seuil, 2010.

Baudry de Saunier, Louis. *Histoire générale de la vélocipédie*. 4th ed. Paris: P. Ollendorff, 1891.

Bellec, François. *Les Sauveteurs: Histoire folle et raisonnée du sauvetage en mer*. Douarnenez: Chasse-Marée, 2008.

Bellencontre, Élie. *Hygiène du vélocipède*. Paris: L. Richard, 1869.

Belloc, Alexis. *La Télégraphie historique: Depuis les temps les plus reculés jusqu'à nos jours*. 2nd ed. Paris: Frimin-Didot, 1894.

Benedict. "Chronique musicale." *Le Figaro*, October 11, 1868.

Benezech, A. "Revue des théâtres." *L'Indépendant français*, January 31, 1869.

Berruyer, Alfred. *Caducêtres ou jambes étrières brevetées s. g. d. g. pour vélocipèdes*. Grenoble: F. Allier père & fils, 1869.

———. *Le Manuel du véloceman*. Grenoble: Prudhomme, 1869.

Besse, Nadine, and Ann Henry, eds. *Le Vélocipède: Objet de modernité*. Saint-Etienne: Musée d'art et d'industrie, 2008.

Bienvenu, Léon. "Fantaisie." *L'Eclipse*, February 16, 1868.

———. "Fantaisie." *L'Eclipse*, August 22, 1869.

Blondeau, Henri. *Dagobert et son vélocipède*. Lyrics by Frédéric Demarquette. Paris: Ch. Grou, 1869.

Bouchery, E. "Bulletin." *La Patrie*, April 7, 1868.

Bouillet, Marie-Nicolas. *Dictionnaire universel d'histoire et de géographie*. Paris: Hachette, 1863.

Brohm, Jean-Marie. *Sport: A Prison of Measured Time*. Translated by Ian Fraser. London: Ink Links, 1978.

Buffon, Georges-Louis de. *Discours sur le style et autres discours académiques*. Paris: Hachette, 1843.

Burr, Thomas. "Markets as Producers and Consumers: The French and United States National Bicycle Markets, 1875–1910." PhD diss., University of California–Davis, 2005.

Céas, Abel. "Le Vélocipède." *Courrier de la Drôme et de l'Ardèche*, December 18, 1867.

Cervantes Saavedra, Miguel de. *The History of Don Quixote de la Mancha*. Hartford, CT: Silas Andrus and Son, 1851.

———. *L'Ingénieux Hidalgo Don Quichotte de la Manche*. Translated and with notes by Louis Viardot. Paris: J.-J. Dubochet, 1836.

Clairville (pseud. of Louis-François-Marie Nicolaïe) and Paul Siraudin. *Le Mot de la fin: Petite Revue en un acte et deux tableaux*. Paris: Michel Lévy Frères, 1869.

Cogniard, Théodore, and Hippolyte Cogniard. *La Chatte blanche*. Paris: Michel Lévy Frères, 1852.

Cohen, Margaret. "Panoramic Literature and the Invention of Everyday Genres." In *Cinema and the Invention of Modern Life*, edited by Leo Charney and Barbara Schwartz, 227–52. Berkeley: University of California Press, 1995.

Coulanges, Louis de. "Les Préfets de la république." *Le Figaro*, March 9, 1872.

"Course de fonds de Paris à Rouen." *Le Vélocipède illustré*, October 7, 1869.

Crestin-Billet, Frédérique. *Veuve Clicquot: La Grande Dame de la Champagne*. Translated by Carole Fahy. Grenoble: Glénat, 1992.

Cropper, Corry. "Like a Furnace: Alfred Jarry's *The Supermale*, Doping, and the Limits of Positivism." In *Culture on Two Wheels: The Bicycle in Literature and Film*, edited by Jeremy Withers and Daniel P. Shea, 94–115. Lincoln: University of Nebraska Press, 2016.

———. *Playing at Monarchy: Sport as Metaphor in Nineteenth-Century France*. Lincoln: University of Nebraska Press, 2008.

C. V. "Lettre d'un convalescent, sur les travaux et embellissemens [*sic*] de Paris." *Journal des artistes*, September 7, 1828.

D . . . , Charles. "Journal d'un véloceman pendant le siège de Paris." *La Vitesse*, July 23 and July 30, 1871.

Dauncey, Hugh. *French Cycling: A Social and Cultural History*. Liverpool: Liverpool University Press, 2012.

Dauncey, Hugh, and Geoff Hare. *The Tour de France, 1903–2003*. London: Frank Cass, 2003.

Delpit, Albert. "La Liberté des théâtres." *Revue des deux mondes* 25 (1878): 601–23.

Delvau, Alfred. *Dictionnaire érotique moderne par un professeur de langue verte.* Basel: Karl Schmidt, 1864.

———. *Les Heures parisiennes.* Paris: Librairie Centrale, 1866.

———. *Les Heures parisiennes.* 2nd ed. Paris: C. Marpon et E. Flammarion, 1882.

Deschamps, Julien. "Théâtre Molière." *Théâtre-Journal: Musique, littérature, beaux-arts,* January 10, 1869.

Deschamps, Théophile. "Echos de Paris et des théâtres." *L'Indépendance dramatique,* January 6, 1869.

Duckett, M. W., ed. *Dictionnaire de la conversation et de la lecture.* Paris: M. Lévy Frères, 1853–60.

Duplessy, J. *Le Guide indispensable des voyageurs sur les chemins de fer de l'Alsace.* Strasbourg: V. Levrault, 1842.

E. P. "Course de vélocipèdes à Saint Cloud." *Le Petit Journal,* June 2, 1868.

"Faits divers." *La Vitesse,* July 23, 1871.

Faulquemont, P. de. "A Travers Paris." *L'Indépendance dramatique,* November 25, 1868.

———. "Chronique de Paris." *L'Indépendance dramatique,* September 23, 1868.

Favre, Alexis-Georges. *Le Vélocipède, sa structure, ses accessoires indispensables, le moyen d'apprendre à s'en servir en une heure.* Marseille: Barlatier-Feissat père et fils, 1868.

Ferry, Paul. "Petits-Bouffes Saint-Antoine." *La Comédie,* September 6, 1868.

———. "Théâtres de Paris." *La Comédie,* August 16 and August 23, 1868 (nos. 296–97).

Flan, Alexandre. "Les Vélocipèdes." *La Chanson illustrée,* May 2, 1869.

Fouillée, Augustine. *Le Tour de la France par deux enfants.* Paris: Belin, 1877.

Fouquier, Henry. "Courrier de Paris." *Le Charivari,* September 17, 1868.

Fournier, Edouard. *Le Vieux-Neuf: Histoire ancienne des inventions et découvertes modernes.* Vol. 2. Paris: E. Dentu, 1877.

G. d'O (possibly a pseud. of Eugène Chapus). "En véloce! En véloce!" *Le Sport, Journal des gens du monde,* July 28, 1867.

Grand-Cateret, John. *Les Mœurs et la caricature en France.* Paris: Librairie illustrée, 1888.

Grenier, Nicolas. *La Petite Reine: Une anthologie littéraire du cyclisme.* Le Crest: Les Editions du Volcan, 2017.

Guizot, François. *Nouveau dictionnaire universel des synonymes.* Paris: Aimé Payen, 1848.

Hadland, Tony, and Hans-Erhard Lessing. *Bicycle Design: An Illustrated History.* Cambridge, MA: MIT Press, 2014.

Herlihy, David. *Bicycle: The History.* New Haven: Yale University Press, 2004.

Horace. *L'Art poétique d'Horace.* Translated by Jacques Peletier du Mans. Paris: M. Vascosan, 1545.

Houssaye, Arsène. *Histoire de Léonard de Vinci.* Paris: Dentu, 1869.

Jehan de la Pédale. *Pédalons!* Paris: Bureaux du "Véloce-Sport," 1892.

Jeux de tous les âges au château de Robert mon oncle, Les Paris: Librairie du Petit Journal, 1867.

Klejman, Laurence, and Florence Rochefort. *L'Egalité en marche: Le Féminisme sous la troisième république*. Paris: Des Femmes, 1989.

Kobayashi, Keizo. *Histoire du vélocipède de Drais à Michaux, 1817–1870: Mythes et réalités*, Tokyo: Bicycle Culture Centre, 1990.

La Chanson illustrée, May 2, 1869.

La Mettrie, Julien Offray de. *L'Homme machine*. Paris: Elie Luzac fils, 1748.

Lamon, Rémy. *Théorie vélocipédique et pratique*. Paris: Imprimerie Nouvelle, 1872.

Larousse, Pierre. *Grand Dictionnaire universel du XIXe siècle*. Vol. 13. Paris: Editions Larousse, 1875.

Lawrence, Scotford. "Paintings of the Velocipede." In *Le Vélocipède: Objet de modernité*, edited by Nadine Besse and Ann Henry, 91–113. Saint-Etienne: Musée d'art et d'industrie, 2008.

———. *The Velocipede: Three Contemporary French Texts*. Cheltenham: John Pinkerton Memorial Publishing Fund, 2014.

Le Bon Roi Dagobert: Chanson ancienne. Paris: A. Huré, 1863.

Le Cousin Jacques (pseud. of Richard Lesclide). "A vélocipède! Messieurs, à vélocipède!" *L'Eclipse*, May 16, 1869.

Le Grand Jacques (pseud. of Richard Lesclide). *Almanach des Vélocipèdes illustré par un cheval sans ouvrage*. Paris: Librairie du Petit Journal, 1869.

———. "A nos amis." *Le Vélocipède illustré*, May 2, 1872.

———. "Faits divers." *La Vitesse*, 16 July 1871.

———. *Le Premier Duel de Pierrot*. Paris: Librairie de l'eau-forte, 1876.

———. *Le Tour du monde en vélocipède*. Paris: Aux Bureaux de la Publication, 1870.

———. *Manuel du vélocipède*. Paris: Librairie du Petit Journal, 1869.

———. *Pierrot en prison*. Paris: Librairie de l'eau-forte, 1876.

Le Numéro 445. "Courrier de pélagie." *Le Figaro*, June 8, 1870.

Le Sage. *Recueil des plus beaux secrets des grands guérisseurs*. Paris: Paul Leymaire, 1931.

Lesclide, Richard. *Contes extragalants*. Paris: E. Dentu, 1886.

———. *La Diligence de Lyon*. Brussels: Henry Kistemaeckers, 1882.

———. *La Femme impossible*. Paris: E. Dentu, 1883.

———. *Une maison de fous*. Paris: Librairie de l'eau-forte, 1876.

Lessing, Hans-Erhard. "What Led to the Invention of the Early Bicycle?" In *Cycle History 11: Proceedings of the 11th International Cycling History Conference*, edited by Andrew Ritchie and Rob van der Plas, 28–36. San Francisco: Van der Plas Publications, 2001.

Leterrier, E., A. Vanloo, and Laurent de Rillé. *Le Petit Poucet: Partition piano et chant*. Paris: Colombier, 1868.

Lindaman, Dana Kristofor. *Becoming French: Mapping the Geographies of French Identity, 1871–1914*. Evanston, IL: Northwestern University Press, 2016.

Lloyd, Rosemary. "Reinventing Pegasus: Bicycles and the Fin-de-Siècle Imagination." *Dix-Neuf* 4, no. 1 (2005): 52–60.

Lovelace, Richard. *Lucasta*. London: John Russell Smith, 1864.

Lugli, Alessandro, Inti Zlobec, Gad Singer, Andrea Kopp Lugli, Luigi M. Terracciano, and Robert M. Genta. "Napoleon Bonaparte's Gastric Cancer: A

Clinicopathologic Approach to Staging, Pathogenesis, and Etiology." *Nature Clinical Practice Gastroenterology and Hepatology* 4 (January 2007): 52–57.

L'Univers illustré. June 6, 1868.

Macy, Sue. *Wheels of Change: How Women Rode the Bicycle to Freedom*. Washington, DC: National Geographic, 2011.

Marks, Patricia. *Bicycles, Bangs, and Bloomers: The New Woman in the Popular Press*. Lexington: University Press of Kentucky, 1990.

Mauté, Mathilde [L'Ex-Madame Paul Verlaine]. *Mémoires de ma vie*. Edited by Michael Pakenham. Seyssel: Champ Vallon, 1992.

McCormick, John. *Popular Theaters of Nineteenth-Century France*. New York: Routledge, 1993.

Merlé, A. "La Fête mâconnaise." *La Petite Presse*, May 19, 1869.

Mesch, Rachel. *Before Trans: Three Gender Stories from Nineteenth-Century France*. Stanford: Stanford University Press, 2020.

Meyland, G. "L'Histoire des sports." *Le Radical*, October 4, 1904

Monréal, Hector, and Henri Blondeau. *Frou-Frou*. Paris: Albert Petit, 1898.

Monselet, Charles. *Figures parisiennes*. Paris: Jules Dagneau, 1854.

Monsieur de la Rue. "L'Insubmersible: Appareil de sauvetage, de sport, d'hygiène, de plaisir et d'utilités diverses." *Le Monde illustré*, March 27, 1869.

Moureau, Jules. "Nouvelles diverses." *Journal de la ville de Saint Quentin*, June 19, 1868.

Muir, Edward. *Ritual in Early Modern Europe*. Cambridge: Cambridge University Press, 1997.

Nettleton, Claire. "Driving Us Crazy: Fast Cars, Madness, and the Avant-Garde in Octave Mirbeau's *La 628-E8*." *Nineteenth-Century French Studies* 42, nos. 3–4 (2014): 250–63.

Nord, Philip. *The Republican Moment: Struggles for Democracy in Nineteenth-Century France*. Cambridge, MA: Harvard University Press, 1995.

"Note sur le vélocipède." *Le Dartagnan*, no. 45, May 16, 1868.

Note sur le vélocipède à pédales et à frein de M. Michaux par un amateur. Paris: Imprimerie de Ad. Laine et J. Havard, 1868.

"Nouvelles des théâtres." *Théâtre-Journal: Musique, littérature, beaux-arts*. July 5, 1868.

Nye, Edward. *A bicyclette*. Paris: Les Belles Lettres, 2000, 2nd ed., 2013.

Olivier, Marc. "Civilization Inoculated: Nostalgia and the Marketing of Emerging Technologies." *Journal of Popular Culture* 44, no. 1 (2011): 134–57.

Olmsted, William. *The Censorship Effect: Baudelaire, Flaubert, and the Formation of French Modernism*. Oxford: Oxford University Press, 2016.

O'Malley, John. *Vatican I: The Council and the Making of the Ultramontane Church*. Cambridge, MA: Harvard University Press, 2018.

"Paris." *Le Petit Journal*, October 30, 1867.

"Paris." *Le Petit Journal*, May 31, 1868.

"Paris au trait." *Paris-Caprice: Revue féerique de l'année*, 1868.

Paz, Eugène. "De l'hydrothérapie." *La Bicyclette*, July 7, 1893.

———. *La Santé de l'esprit et du corps par la gymnastique: Etude sur les exercices du corps depuis les temps les plus reculés jusqu'à nos jours, leurs progrès, leurs effets merveilleux, leurs diverses applications et leur combinaison avec l'hydrotérapie* [*sic*]. Paris: La Librairie du Petit Journal, 1865.

Peloq, Jules. "Une course de vélocipèdes." *L'Univers illustré*, June 6, 1868.

Perrodil, Édouard de. "La Bicyclette." *Les Echos*. Paris: Vanier, 1891.

Pétis de la Croix, François. *Mille et un jours*. 5 vols. Paris: Béchet Aîné, 1710–12.

Petit, Léon. "La Bicyclette et le corset." In *Paris-vélo almanach*, 61. Paris: G. Charpentier and E. Fasquelle, 1896.

Picard, Alfred. *Les Chemins de fer: Aperçu historique, résultats généraux de l'ouverture des chemins de fer, concurrence des voies ferrées entre elles et avec la navigation*. Paris: H. Dunod et E. Pinat, 1918.

Pierre et Paul. "Notes et souvenirs." In *Le Roannais illustré*. Series 6. Roanne: n.p., 1892.

Pneumatic. "Choses de sport." *Le Travailleur normand*, April 19, 1896.

Pourielle, Edgard. "Théâtre de l'Athénée, *Le Petit Poucet*." *Le Théâtre illustré*, October 1868.

Poyer, Alex. *Les Premiers Temps des véloce-clubs*. Paris: L'Harmattan, 2003.

———. "The Origins of Velocipede Clubs." In *Le Vélocipède: Objet de modernité*, edited by Nadine Besse and Ann Henry, 77–83. Saint-Etienne: Musée d'art et d'industrie, 2008.

Prével, Jules. "Petit Courier des théâtres." *Le Figaro*, September 3, 1868.

Programme de la fête nationale du 15 août 1855. Paris: Gaittet et Cie, 1855.

Rabelais, François. *Œuvres de maître François Rabelais publié sous le titre de Faits et dits du géant Gargantua et de son fils Pantagruel*. Amsterdam: Henri Bordesius, 1711.

Rancière, Jacques. *The Ignorant Schoolmaster: Five Lessons in Intellectual Emancipation*. Palo Alto: Stanford University Press, 1991.

Ravenel (attrib.). "Pour le former à la souplesse . . ." In *Histoire générale de la vélocipédie*, by Louis Baudry de Saunier, 4th ed., 47–48. Paris: Paul Ollendorff, 1891.

Rieux, Jules de, and Charles d'Orvict. *Au restaurant de la Porte Jaune*. Paris: Le Bailly, 1878.

Riol, Jules. *La Bicyclette, monologues en vers, dédié au Touring Club de France*. Paris: Lanée, 1896.

Ritchie, Andrew. *Early Bicycles and the Quest for Speed: A History, 1868–1903*. 2nd ed. Jefferson, NC: McFarland, 2018.

Rivarol, René. "Entre deux bourses." *Le Figaro*, September 16, 1868.

Robin, Francis. *Le Paris-Versailles du dimanche 8 décembre 1867: Première Course cycliste de tous les temps?* Pomeys: Vélocithèque, 2017.

Rouvrelle. "Menus-Plaisirs." *La Comédie*, January 3, 1869.

R. T. "A Monsieur Louis Veuillot." *Le Vélocipède illustré*, April 1, 1869

Sacerdot, Max. "Menus-Plaisirs." *Le Théâtre illustré*, January 1869.

Schwartz, Vanessa. *Spectacular Realities: Early Mass Culture in Fin-de-Siècle Paris*. Berkeley: University of California Press, 1998,

Seray, Jacques. *Richard Lesclide: Du Vélocipède illustré à la table de Victor Hugo*. Vélizy: J. Seray, 2009.

Sieburth, Richard. "Same Difference: The French Physiologies, 1840–1842." In *Notebooks in Cultural Analysis: An Annual Review*, edited by Norman F. Cantor, 163–200. Durham, NC: Duke University Press, 1984.

Six, Alexandre. "Les Vélocipèdes." Lille: Imprimerie de Six Horcmans, 1869.

Smethurst, Paul. *The Bicycle: Towards a Global History.* London: Palgrave Macmillan, 2015.

Stel, Adolphe. "Premières représentations." *L'Indépendance dramatique*, September 30, 1868.

Suérus, Raoul. *Le Vélocipède: Vers lus au banquet de la Saint-Charlemagne.* Paris: Lycée Impérial Saint-Louis, 1869.

Thackeray, William Makepeace. *The Paris Sketch Book.* New York: D. Appleton, 1852.

"Théâtres." *La Presse.* November 22, 1869.

Thiéry, Henri. *Les Contributions indirectes.* Paris: Dentu, 1866.

Thompson, Christopher S. *The Tour de France: A Cultural History.* Berkeley: University of California Press, 2008.

Touchatout (pseud. of Léon Bienvenu). "Vélocipède IV." *Le Trombinoscope*, no. 111, 1873.

Trimm, Timothée. "Les Vélocipèdes." *Le Petit Journal*, July 5, 1868.

Union vélocipédique de France. *Bulletin officiel*, 1907.

Vabontrain (pseud. of Philippe Dubois). *Manuel de santé tintamarresque du docteur Vabontrain, ou Cours d'idiopathie à l'usage des gens du monde.* Paris: Impr. de E. Mazereau, l'an des vélocipèdes, 1869

Varro, Marcus Terrentius. *M. Terentii Varronis Saturarum Menippearum Reliquiae.* Edited by A. Riese. Berlin: Teubner, 1865.

Vautier, A. "Chronique." *L'Indépendance parisienne*, October 16, 1868.

Verlaine, Paul. "Dans la grotte." In *Fêtes galantes*, 11–12. Paris: Alphonse Lemerre, 1869.

———. *Œuvres poétiques complètes.* Paris: Robert Laffont, 1992.

Verne, Jules. *Voyages et aventures du capitaine Hatteras.* Paris: Hetzel, 1866.

Veuillot, Louis. *Les Couleuvres.* Paris: V. Palmé, 1869.

Vigne, Henri. "Le Vélocipède." *L'Illustration: Journal universel* 53, 1869.

Villette, Léon de. "Nouvelles locales et faits divers." *L'Industriel de Saint-Germain-en-Laye*, May 9, 1868.

Weber, Eugen. *France, Fin de Siècle.* Cambridge, MA: Harvard University Press, 1986.

Wetzel, David. *A Duel of Giants: Bismarck, Napoleon III, and the Origins of the Franco-Prussian War.* Madison: University of Wisconsin Press, 2003.

Willis, Henry. *A History of the Latin Monetary Union: A Study of International Monetary Action.* Chicago: University of Chicago Press, 1901. Reprint, New York: Greenwood Press, 1968.

Wils, Nicole. *Dictionnaire des théâtres parisiens au XIXe siècle.* Paris: Aux Amateurs de Livre, 1989.

X. "Petites nouvelles." *Le Gaulois*, September 21, 1868.

Yriarte, Charles. "Courrier de Paris." *Le Monde illustré*, June 26, 1869.

Zola, Émile. "La Littérature et la gymnastique." *Salut public*, October 5, 1865.

ILLUSTRATION CREDITS

INTRODUCTION

Figures I.1–I.6. gallica.bnf.fr / Bibliothèque nationale de France (National Library of France)

Figure I.7. Universitätsbibliothek Heidelberg (Heidelberg University Library)

Figure I.8. gallica.bnf.fr / Bibliothèque nationale de France

Figure I.9. Cambridge University Library

CHAPTER ONE

Figure 1.1. gallica.bnf.fr / Bibliothèque nationale de France

Figure 1.2. RMN-Grand Palais / Art Resource, NY. Photo: Gérard Blot

Figure 1.3. Paris Musées: Carnavalet Museum, History of Paris

CHAPTER TWO

Figure 2.1. gallica.bnf.fr / Bibliothèque nationale de France

CHAPTER THREE

Figures 3.1–3.2. gallica.bnf.fr / Bibliothèque nationale de France

Figure 3.3. *Manuel du vélocipède* (*Manual of the Velocipede*) (1869) by Le Grand Jacques with illustrations by Emile Benassit, courtesy of University of Michigan Library; digitization provided by Brigham Young University Digital Imaging Lab

Figures 3.4–3.5. Bayerische Staatsbibliothek München (Bavarian State Library Munich), 2 per. 15 s-3, Scan 167, urn: nbn: de: bvb: 12-bsb10498606-5

Figures 3.6–3.28. *Manuel du vélocipède* (*Manual of the Velocipede*) (1869) by Le Grand Jacques with illustrations by Emile Benassit, courtesy of University of Michigan Library; digitization provided by Brigham Young University Digital Imaging Lab

Figure 3.29. Bayerische Staatsbibliothek München, 2 per. 15 s-3, Scan 255, urn: nbn: de: bvb: 12-bsb10498606-5

Figures 3.30–3.31. *Manuel du vélocipède* (*Manual of the Velocipede*) (1869) by Le Grand Jacques with illustrations by Emile Benassit, courtesy of University of Michigan Library; digitization provided by Brigham Young University Digital Imaging Lab

CONCLUSION

Figures C.1–C.2. gallica.bnf.fr / Bibliothèque nationale de France

INDEX

Page numbers in italics refer to illustrations.

ABOUT THE AUTHORS

Corry Cropper is a professor of French at Brigham Young University. He is the author of *Marianne Meets the Mormons: Representations of Mormonism in Nineteenth-Century France* (with Heather Belnap and Daryl Lee), *Mormons in Paris: Polygamy on the French Stage, 1874–1892* (with Christopher M. Flood, Bucknell University Press), and *Playing at Monarchy: Sport as Metaphor in Nineteenth-Century France.* When he is not teaching or writing, he can be found cycling in the mountains around Provo, Utah.

Seth Whidden is a professor of French literature and fellow of the Queen's College at the University of Oxford in the United Kingdom, and the editor of *Nineteenth-Century French Studies*. His publications include *Leaving Parnassus: The Lyric Subject in Verlaine and Rimbaud*, *Authority in Crisis in French Literature, 1850–1880*, and *Reading Baudelaire's "Le Spleen de Paris" and the Nineteenth-Century Prose Poem.* He divides his time between Oxford's bike lanes and cobblestones and the dirt roads of Cliff Island, Maine.